30-SECOND
ZOOLOGY

30-SECOND ZOOLOGY

The 50 most fundamental
categories and concepts
from the study of animal life

Editor
Mark Fellowes

Contributors
James Barnett
Amanda Callaghan
Peter Capainolo
Mark Fellowes
Neil Gostling
Rebecca Thomas

Illustrations
Nicky Ackland-Snow

IVY PRESS

First published in the UK in 2020 by
Ivy Press
An imprint of The Quarto Group
The Old Brewery, 6 Blundell Street
London N7 9BH, United Kingdom
T (0)20 7700 6700
www.QuartoKnows.com

British Library Cataloguing-in-
Publication Data
A catalogue record for this
book is available from the
British Library.

ISBN: 978-1-78240-853-6

This book was conceived,
designed and produced by
Ivy Press
58 West Street, Brighton BN1 2RA, UK

Publisher **David Breuer**
Editorial Director **Tom Kitch**
Art Director **James Lawrence**
Project Editor **Stephanie Evans**
Designer **Ginny Zeal**
Illustrator **Nicky Ackland-Snow**
Editorial Assistant **Niamh Jones**

Printed in China

10 9 8 7 6 5 4 3 2 1

CONTENTS

INTRODUCTION
Mark Fellowes

Zoology, the study of animal life in all its forms and structures, is a subject that has captivated people for millennia. In part, this is for simple, practical reasons: animals are a source of food and materials, and some of those we have domesticated also provide us with their labour and companionship. Of course, other, wild, animals can also threaten us with physical harm and disease. But animals are much more than this; they are an innate part of our own species' psyche. In parts of France you can see cave paintings executed over 15,000 years ago that depict species such as horses, bison and ibex in such detail that even today they can be readily identified. But this is not something unusual; everywhere across the world where we find ancient art we see representations of animals, and this desire to identify with other species remains a part of our culture.

The science of zoology has a long history, but we can think of it as having two phases: what happened before Darwin and the exponential growth in knowledge that occurred after him. The earliest surviving systematic studies of animals were by the Greek philosopher Aristotle (384 BCE–322 BCE). While living on the island of Lesbos he studied the diversity of animal life around him, and he introduced key ideas that influenced thinkers for many centuries, what we now think of as metabolism, temperature regulation, sensation processing, embryonic development and inheritance.

It was not until the Renaissance that Aristotle's work was truly built upon, with a renewed focus on the importance of empirical observations in drawing conclusions about the natural world. This flowered after the publication of Conrad Gessner's *Historia Animalium* in the 1550s, which was the first attempt to describe all known animal species. The number of people studying animals grew, and taxonomists, such as Carl Linnaeus (1707–78), attempted to bring coherence to our understanding of the diversity of life, and great natural historians, among them John James Audubon (1785–1851), signicantly expanded our knowledge of the animal kingdom as they explored the world.

But it is the naturalist Charles Darwin (1809–81) who truly changed zoology, just as he changed biology and beyond. In *On the Origin of Species* (1859), he showed not only that species were not unchanging but, more importantly, he provided a mechanism that explained how that change could occur though natural (and later sexual) selection. Zoology's explanations for many of the wondrous forms and behaviours of animals are based on his insights, and the great explosion in zoological knowledge that followed Darwin truly placed our species' history as being one among the animals rather than being somehow separate from the rest of nature.

Zoology is not an arcane science; it is highly accessible, and an understanding of zoology is central to our future. Zoologists apply themselves to some of the most important challenges facing society today: they study diseases that kill millions, such as malaria; they investigate how to control crop pests, such as locusts, that can devastate the poorest parts of the world; they help us to develop ways to use resources, for example fisheries, sustainably; they develop ecotourism projects to protect the last remaining great apes. But zoologists also ask questions about the world and our place in it. Why is biodiversity highest in the tropics? How do birds migrate with precision halfway around the world? Why are some groups so rich in species, whereas others only have one or two? Why do some animals use tools and others do not? The former approach addresses urgent threats, while the latter satisfies our innate curiosity.

In this book we have focused on what we consider to be the 50 most important concepts in zoology. Our aim is to we show just how broad, fascinating and relevant the subject is as a science for the world today. We start by considering the **Origin & Evolution** of animals, explaining how the exuberant diversity of animal life came into being and how an understanding of this helps explain some strange observations we see in isolated islands. In the following two chapters we review the best known of the major animal groups, **The Invertebrates** and **The Vertebrates**. Then, in Chapter 4, we consider elements of **Physiology**, how animal bodies work to allow them to adapt to different environments, covering key concepts such as breathing and flight. In Chapter 5 we review key concepts in **Behaviour**, that remarkable ability animals have to respond to changes in their environments and to each other to enhance their chances of surviving and reproducing. In the penultimate chapter we consider **Ecology**, the science of how animals interact with each other and their environments, determining their abundance and distributions. It is only through an understanding of animal ecology that we can hope to mitigate the challenges faced by animals in coming decades, and this is the topic of our final chapter, **Conservation & Extinction**. Many zoologists would agree that we are living through one of the greatest mass extinctions seen in the history of the Earth, and there is little controversy over the realization that the activities of humankind are the root cause. We over-exploit and under-protect the great diversity of life on our planet, and if we are to have any hope of protecting the most awe-inspiring of species in their natural habitats, then we must act now. It is only by accepting our own complicity in this devastation and then acting to mitigate it that we can hope to ensure that future generations are as astonished by the grandeur of animal life as we have been.

Life is truly precious. Let us resolve together to learn more about these astonishing creatures, to love our environment and the myriad animals it sustains, and to protect what remains for the generations who will follow us.

ORIGIN & EVOLUTION

analogous Referring to a characteristic derived by convergent evolution to solve a problem, but not sharing evolutionary history, such as the wings of birds and insects.

bases and triplet code Specifically genetic terms: DNA is composed of nucleotide bases A, C, T and G (U replaces T in RNA). Bases in triplets (groups of three) specify the amino acid building blocks of proteins.

bilaterians Animals with three body axes: left–right, dorsal–ventral (back–front) and anterior–posterior (head–tail).

central dogma (of molecular biology) The DNA in the nucleus is copied, making a molecule called RNA that provides the molecular instructions to make proteins.

choanocyte The cells within a sponge that have a beating flagellum surrounded by a collar (Gr. *choana*) that draws water and food particles through the sponge.

choanoflagellates A group of single-celled organisms that resemble isolated choanocytes of sponges. They are the closest related single-celled organism to animals.

clade A collection of organisms that group together through the possession of shared characteristics inherited from a common ancestor. Birds are an example of a clade.

conodonts Eel-like marine vertebrates.

convergent evolution The process that allows common problems to be solved in similar ways, independently of evolutionary relatedness. Flight in birds and bats has used convergent evolution to independently turn the arm into a wing in each group.

DNA The molecule that provides the chemical instructions for making individual organisms. DNA comprises four molecules: Adenine – A, Cytosine – C, Thymine – T and Guanine – G. The order of letters provided the DNA code.

eukaryotic The group of organisms that have a defined nucleus that contains the DNA homology/homologous: similar characteristics shared because of inheritance from a common ancestor, such as the forelimb in vertebrates.

flagellum (pl. flagella) A whip-like beating structure, extending from cells, often enabling locomotion.

genome duplication An episode in tetrapod history, when all of the genes were doubled up, and then the entire double set was doubled from 1 to 2 to 4 copies of genes found in invertebrates.

geological time A measure in eras (for example, Palaeozoic, Mesozoic) and periods (for example, Triassic, Jurassic, Cretaceous) and defined by the fossils found in them. The changes in fossils, the extinction events themselves, bracket geological time periods.

heterotroph An organism that obtains its nutrition from 'eating' organic material.

homologous Similar characteristics shared because of inheritance from a common ancestor, such as the forelimb in vertebrates.

industrial melanism Term to describe the response that saw melanic (blacker) varieties of species increase in response to pollution causing a darkening of habitats, such as the trunks of trees.

K-T ME Abbreviation for the Cretaceous-Tertiary Mass Extinction event, approximately 66 million years ago, that saw the demise of the dinosaurs.

monophyletic group Organisms that included the last common ancestor and all of its descendants, living and extinct.

niche The set of ecological requirements of a species or population; it covers all of the aspects that a species needs to survive.

phylum (pl. phyla) The first taxonomic division of animals below the kingdom.

protist A single-celled organism. There are protist partners to all the multicellular kingdoms: Plantae, Animalia and Fungi.

RNA The intermediary molecule that is the same sequence as DNA (see opposite), which provides the code that is translated into a protein. RNA substitutes Thymine for Uracil, and therefore has a 'U' in place of 'T'.

teleost A bony, ray-finned fish, such as the cod or goldfish.

tetrapod A four-limbed vertebrate.

transcription The act of copying the 'ACTG' code of DNA into the 'ACUG' code of RNA.

translation The act of changing from the 'ACUG' 'triplet code' of RNA into the amino acid sequence in a protein.

8 January 1823
Born in Llanbadoc, Usk,
South Wales, to Thomas
and Mary Wallace

1828
Family moves to
Hertford, England

1837
Wallace moves to London
to train with his brother
to become a surveyor

1844
Takes up a teaching post
in Leicester, where he
meets Henry Walter
Bates, an entomologist

1848
Aged 25, Wallace
accompanies Bates
aboard the *Mischief* for
an expedition to Brazil
where he independently
collects many thousands
of animal specimens

1852
Travels back to England
on board the *Helen* but
the vessel catches fire
and sinks. Most of
Wallace's specimens,
drawings and field notes
are lost

1854
Embarks on an expedition
to the Malay Archipelago

18 June 1858
Sends his paper on
natural selection to
Charles Darwin

1 July 1858
Darwin and Wallace's
work is presented to the
Linnaean Society, a year
before Darwin's own
masterpiece is published

1869
Wallace publishes *The
Malay Archipelago*, an
account of his eight years
in the region

1876
Publishes *The
Geographical Distribution
of Animals*, which drew
on his identification of
distinct biogeographical
zones

1881
Awarded an annual
pension by the British
government

7 November 1913
Dies and is buried in
Broadstone, Dorset

ALFRED RUSSEL WALLACE

Alfred Russel Wallace was an unlikely character to rise to such prominence in the scientific world of the nineteenth century. Born in what is now Monmouthshire, South Wales, in 1823, the son of a solicitor of modest means, he left school at the age of 14 and followed his older brother to London. There he trained as a surveyor, but other interests drew his attention, and he became an amateur botanist. By the early 1840s he had left the surveying business and spent a year in Leicester as a teacher. He briefly returned to surveying, but his passion for what would become the science of biology was developing in earnest.

Socially and politically Wallace was influenced by the Welsh utopian socialist Robert Owen as well as the great scientific writers of the time, including the Prussian polymath Alexander von Humboldt, Charles Darwin and the (originally) anonymous author of the 1844 natural history work *Vestiges of the Natural History of Creation* (later revealed to have been written by the Scottish evolutionary thinker and publisher Robert Chambers).

Wallace's contributions to science began with expeditions to collect exotic animals and describe the biogeography of different parts of the world. He travelled in the Amazon basin (1848–52) and the Malay Archipelago (1854–62), where he collected over 100,000 specimens and described hundreds of new species. It was in the Moluccas that Wallace, suffering from a malarial fever, had his epiphany about why, across all of the islands he visited in Southeast Asia, there were unique species – just as Darwin had done in the Galapagos.

The parallel trajectory of Wallace's and Darwin's thinking later became apparent, and in their writings it can be seen how both men independently came to similar conclusions about the nature of nature. Wallace wrote his ideas down in a paper, *On the Tendency of Varieties to Depart Indefinitely from the Original Type*. This, together with Darwin's *Extract from an Unpublished Work on Species*, was presented to the Linnaean Society of London in 1858.

In 1869 Wallace published his most celebrated book, *The Malay Archipelago*, which described the eight years he had spent in the region. Central to his research was his description of a biogeographical border – Wallace's Line – between the islands of Bali and Lombok. Only 35 kilometres (22 miles) apart, these islands belong to different biogeographical zones with differing evolutionary histories: east are the flora and fauna of Australia; west are those of Asia.

Wallace later withdrew from the public eye, but his contributions to science and natural selection were recognized during his lifetime – and the title of his 1889 volume on evolutionary biology gave us a term that lives with us today: Darwinism.

Neil Gostling

GENES

the 30-second zoology

Genes are the DNA instructions that produce organisms. Invertebrates have roughly 10,000 genes, while tetrapod vertebrates have about 25,000. This provides clear evidence of genome duplication, episodes where the number of genes in vertebrate lineages were exponentially increased. This idea was further supported by the discovery of a subset of genes called homeobox genes (Hox for short), where vertebrates have four copies of each Hox gene compared with invertebrates' single copy of each. Hox genes express the head-to-tail plan of all bilaterians (animals with three body axes) during embryonic development and indicate to cells their position in the body so that regions develop the correct structures. Tetrapods have four times as many Hox genes as invertebrates. However, gene number alone does not explain why we are different from other animals. Humans have the same number of genes as chimps, sharks, chickens and snakes, whereas teleost fish, such as cod or zebrafish, have twice as many as we do. How genes are regulated, and the complexity of gene interactions, is far more important in determining the complexity of the resulting organism than simply the number of genes an organism has.

3-SECOND DISSECTION
The 'central dogma' of genetics is DNA → RNA → protein: a gene is transcribed into RNA, which is translated into a protein molecule that performs a function within the cells of living organisms.

3-MINUTE SYNTHESIS
Hox genes, and their significance to the body plan of organisms, were first discovered in the fruit fly, *Drosophila*. If Hox genes are expressed in the wrong place, the wrong appendage forms. This occasionally happens naturally, and fruit flies can develop legs where antennae should be. In tetrapod vertebrates, the increase in Hox gene number provided the genetic 'toolkit' for fins to evolve into limbs and to pattern the pentadactyl hands that were used to type these words.

RELATED TOPICS
See also
NATURAL SELECTION
page 18

PHYLOGENETICS & DIVERSITY
page 22

30-SECOND TEXT
Neil Gostling

Whether fish, fly or human, there are genes that provide a toolkit form, building all animal bodies.

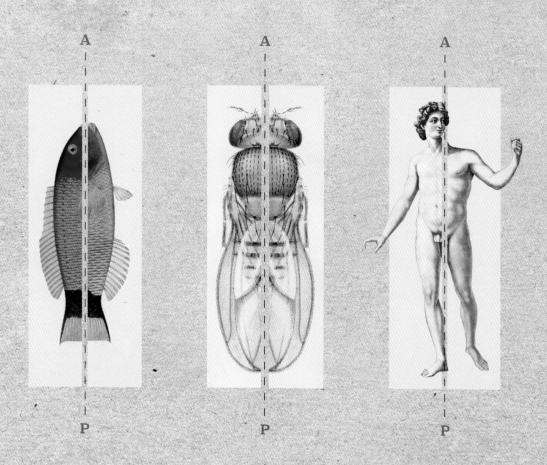

A   A   A

P   P   P

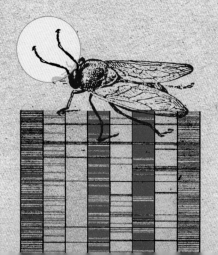

NATURAL SELECTION

the 30-second zoology

Natural selection is one of the easiest mechanisms in science to understand. It is the big idea that sets apart Charles Darwin's writings on evolution from earlier descriptions of species change by scientists and natural philosophers. Evolution by natural selection relies on three simple premises: first, resources are limited in nature so not all individuals can survive to reproduce; second, individuals vary in traits determined by the individual's genetics, which allow them to exploit those resources and survive; third, individuals with those traits are more likely to reproduce and pass on the genes which shaped those survival traits. It is as simple as that. The environment is a filter, determining which individuals, and therefore which traits, make it through to the next generation. As the environment changes, the filter changes, so over time we see species adapt and evolve. One example is industrial melanism. The peppered moth is typically white bodied with dark speckles to camouflage it against tree trunks. With the advent of the Industrial Revolution, pollution covered tree trunks, exposing the moths to bird predation. A rare dark form of the moth did well in these polluted habitats, and it became dominant. As pollution controls were introduced the dark form declined, as the environmental filter of soot-covered trees was removed.

RELATED TOPICS
See also
GENES
page 16

THE FIRST ANIMALS
page 20

3-SECOND BIOGRAPHY
CHARLES DARWIN
1809–82
English naturalist and author of many books, including *On the Origin of Species* (1859); Darwin wrote some of the fundamental works that explain our natural world

30-SECOND TEXT
Neil Gostling

3-SECOND DISSECTION
Individuals best suited to an environment will obtain resources, reach adulthood and be most likely to reproduce, meaning that their characteristics are passed on to the next generation.

3-MINUTE SYNTHESIS
Evolution is the observed reality that life has changed, and natural selection is the mechanism. Today's living organisms have a single common ancestor, which probably evolved about 4 billion years ago. As environments changed, selective pressures drove change in organisms. It might appear that organisms are so well adapted to their environments that there is design and intent, but it is simply the result of less well-adapted organisms not surviving or reproducing as effectively as those with a better fit.

It took a brain like Darwin's to predict, purely on sight of an orchid on Madagascar, that such flowers needed to be fertilized by a pollinator in possession of an extraordinarily long proboscis.

THE FIRST ANIMALS

the 30-second zoology

3-SECOND DISSECTION
Of the roughly 2 million named animal species, around two-thirds are insects (of which half are beetles) and about 50,000 chordates, of which only about 5,000 are mammals.

3-MINUTE SYNTHESIS
Dickinsonia, a 558-million-year-old fossil of a large flat organism – variously described as a fungus and a bacterial mat – has been shown to have been an animal. In 2018 cholesterol was isolated from these fossils and, when analysed, found to be a form that is present only in animals. The Cambrian phyla still exist today, but animals were 'experimenting' with body plans long before an animal left its trackway 542 million years ago at the base of the Cambrian period.

The first animals evolved from their protist relatives, the choanoflagellates, over 700 million years ago. Choanoflagellates (from ancient Greek *choane*, funnel) are peculiar single-celled organisms with a funnel-shaped collar surrounding a flagellum. They are normally solitary but sometimes form aggregations, and they have the cell–cell adhesion and cell-signalling molecules critical to the multicellularity of 'animals'. (Indeed, the choanocytes that line the inside of sponges look almost identical to them, highlighting the close affinity that animals and choanoflagellates share.) As well as being multicellular, animals are mobile heterotrophs, which means that they obtain their nutrition from eating other things. The Cambrian period (542–510 million years ago) was a period of rapid diversification of animals. Most of the phyla – the major animal groups alive today, including the chordates to which the vertebrates (animals with backbones) belong – appear in the Cambrian fossil record. The extraordinary diversification has been put down to an 'arms race', as predators and prey developed more and more sensitive eyes, enabling them to hunt or avoid becoming prey, and skeletons to help in defence. The Cambrian fossil record may, in fact, simply be a result of the evolution of fossilizable – that is, hard-bodied – animals that had evolved some time earlier.

RELATED TOPICS
See also
GENES
page 16

PHYLOGENETICS & DIVERSITY
page 22

CHORDATES
page 56

3-SECOND BIOGRAPHY
CHARLES DOOLITTLE WALCOTT
1850–1927
American palaeontologist who discovered the Burgess Shale Formation in British Columbia, Canada, in 1909; this is one of the earliest fossil beds, remarkable for the preservation of the soft parts of fossils

30-SECOND TEXT
Neil Gostling

By the Cambrian period, 542 million years ago, almost all of the phyla – a huge diversity of animals – were present.

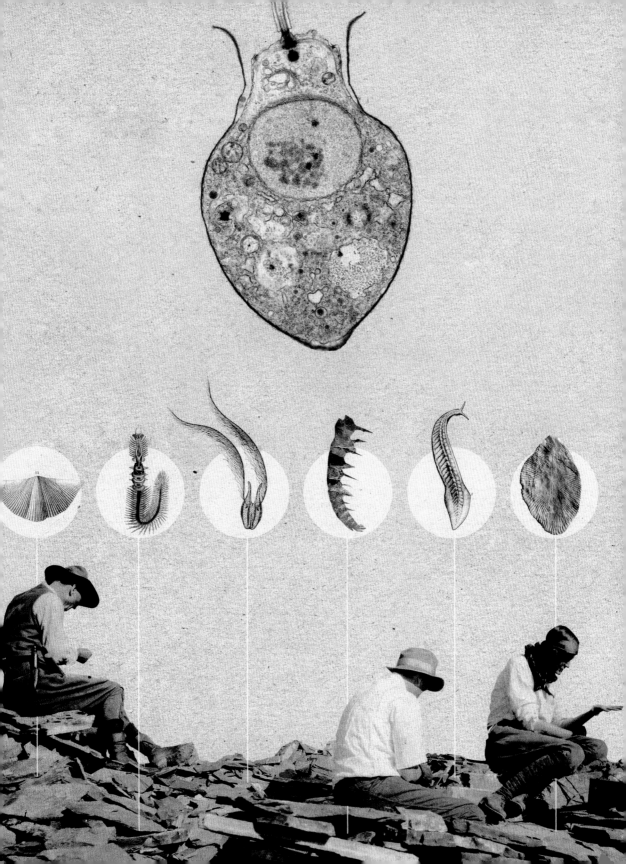

PHYLOGENETICS & DIVERSITY

the 30-second zoology

Estimates suggest that there are around 9 million eukaryotic species, of which two million are animals, over half of them insects. Mammals comprise only 5,000 species and birds 10,000. The system of nomenclature – the naming of all organisms, formalized by the eighteenth-century Swedish scientist Carl Linnaeus – remained the basis of biological classification for over two centuries. His scheme was hierarchical and without evolutionary context; the concept of phylogeny arose with Darwin's theory of evolution. Phylogenies are produced by comparing shared derived, morphological and molecular characteristics and include an ancestor and all descendants, living and extinct. It seeks to classify all groups of organisms according to relatedness and is concerned with clades, groups formed from a common ancestor of which all descendants form monophyletic groups. Phylogenetics abandons some of Linnaeus's hierarchy (species, genus, family, order, class, phylum, kingdom) to reflect evolutionary history. For example, the Linnaean class Reptilia is not monophyletic because the reptiles share their last common ancestor with both birds and mammals. Therefore, Reptilia is not a group. However, this common ancestor, dating back to the Carboniferous period (330 million years ago), was the first animal with an amniotic membrane in its egg.

RELATED TOPICS
See also
NATURAL SELECTION
page 18

THE FIRST ANIMALS
page 20

3-SECOND DISSECTION
Phylogenetics, the study of evolutionary relationships between species, combines genetics, comparative anatomy and mathematics with computing to build species' family trees.

3-MINUTE SYNTHESIS
Phylogenetics is based on shared derived characteristics and develops Darwin's nested 'groups within groups' idea of how species are related. Key to this is recognizing homologous characters, derived by descent with modification. For example, the forelimbs of vertebrates (fish fins, dog legs, human arms and bird wings) are homologous, the result of common ancestry. However, birds' and bats' wings, which have the same function, are analogous, that is, independently derived by convergent evolution, enabling flight.

3-SECOND BIOGRAPHY
CARL LINNAEUS
1707–78
Swedish botanist and author of *Systema Naturae* (1735) who devised the binomial system of classifying plants and animals

WILLI HENNIG
1913–76
The father of systematic phylogenetics; his work helped to bring taxonomy fully into an evolutionary framework, something Linnaean classification lacked

30-SECOND TEXT
Neil Gostling

Birds, mammals and 'reptiles' reproduce by means of an amniotic egg; they are common to the monophyletic group the Amniota, whereas egg-laying fish and amphibians are not.

MASS EXTINCTIONS

the 30-second zoology

Extinction is a fundamental part of evolution. Over the past 500 million years there have been five mass extinctions (MEs) and many smaller events. The first, the Ordovician ME, 450 million years ago, witnessed large declines in marine life with the onset of global cooling. Thereafter, in the Devonian ME (c. 364 million years ago) over half the marine genera – corals, vertebrates, trilobites and ammonites – disappeared, followed by the Permian ME (252 million years ago) when 96 per cent of species died out when widespread volcanic activity caused rapid climate change. Insects suffered their only mass extinction and trilobites became extinct, along with the dominant mammal-like reptiles. Their vacation of niches allowed another group of reptiles to radiate and evolve into the dinosaurs. A further event at the end of the Triassic period (c. 200 million years ago) removed half of all genera, including amphibians, reptiles and the conodonts. However, dinosaurs were able to become the dominant terrestrial group throughout the Mesozoic era, filling all niches and even taking to the skies. When an asteroid struck Earth 66 million years ago, the event, known as the K-T ME, wiped out the dinosaurs and the vast majority of birds. One group, a relic of the Permian, was then able to diversify and become the mammals we know today.

3-SECOND DISSECTION
A mass extinction event is when more than 60 per cent of species disappear in less than a million years.

3-MINUTE SYNTHESIS
On the geological timescale, mass extinctions are 'sudden' events, but there is a natural continuous process known as the background extinction rate. Although difficult to calculate, the average lifespan of a species is 1–4 million years, but human activities are estimated to be hastening the loss of plants and animals to roughly a thousand times the background rate of extinction. The current loss of biodiversity suggests that we are living through the sixth ME.

RELATED TOPICS
See also
CLIMATE CHANGE
page 138

HABITAT LOSS
page 140

HUMAN–WILDLIFE CONFLICT
page 148

3-SECOND BIOGRAPHY
MICHAEL BENTON
1956–
British palaeontologist and author of *When Life Nearly Died*, which describes all aspects of the Permian ME

30-SECOND TEXT
Neil Gostling

Evolutionary history has been punctuated by mass extinctions. The same events that cause some species to become extinct provide opportunities for others to evolve.

ISLAND ODDITIES

the 30-second zoology

Islands do odd things to animals.
Insular life allows populations to exist in splendid isolation from others of their kind, but islands have limited resources, which results in some peculiar evolutionary trajectories. Species that can best take advantage of their environment survive and reproduce. Where the new environment provides a reduced food source or no predators, individuals whose lineage reduces in size can survive on less food or need not invest in growth to escape predation. On Malta, for example, there are fossils of a pygmy elephant that measured less than a metre at the shoulder. While island life tends to make large animals smaller, sometimes small ones become bigger to fill a role. On the Galápagos Islands, where there were no mammals, tortoises became giants and the dominant large herbivore. The dodo of Mauritius was a (very) large pigeon. Like many other birds of isolated islands, it also stopped flying, which, like change in size, is also explained by trade-offs. Flying is very energy intensive. If there is no advantage to flying, then it becomes a hindrance. Individuals that do not invest in flight can invest more in reproduction and so may have an advantage, and the ability to fly will be lost over time. Adapt or die. It's the story of evolution.

3-SECOND DISSECTION
Species respond to environmental conditions or die out. On islands, where resources are limited and predation reduced, the adaptive response is often a radical change in size.

3-MINUTE SYNTHESIS
It's not only animals like elephants, tortoises and dodos that are changed by island life. A lineage of *Homo* (likely *H. erectus*) was isolated on Flores, Indonesia, where they became very small. Discoveries made in 2003 showed that the population of diminutive humans stood little more than a metre tall. Their isolation allowed them to survive until 12,000 years ago, by which time our own species, *H. sapiens*, had outcompeted all other human species across the planet.

RELATED TOPICS
See also
NATURAL SELECTION
page 18

PHYLOGENETICS & DIVERSITY
page 22

30-SECOND TEXT
Neil Gostling

Selection pressures on island-dwelling species over evolutionary time result in dwarfism in some species and gigantism in others.

THE INVERTEBRATES

anthozoa Corals and anemones that have the polyp form of cnidarian body plan.

bilaterians Animals with three body axes: left–right, dorsal–ventral (back–front) and anterior–posterior (head–tail).

biomass The total quantity or weight of organisms in a given area or volume.

biosynthesis The biological process whereby molecules are converted into more complex products, often in a multi-step, enzyme-catalyzed process.

chaete (pl. chaetae) Body hairs or bristle.

chelicerataes The first pair of appendages on the head segment of a chelicerate.

chemoreceptor A sensory cell or organ that can detect chemical stimuli, such as smell or taste, and relay that information to the central nervous system.

chitin The material that forms the exoskeletons of arthropods, the radulae of molluscs, the scales of fish and some amphibians and the beaks of cephalopods like the squid and octopus.

choanocyte The cells within a sponge that have a beating flagellum surrounded by a collar (Greek. *choana*) that draws water and food particles through the sponge.

clade A collection of organisms that group together through the possession of shared characteristics inherited from a common ancestor.

clitellum A thick, saddle-like section of a worm or leech in which eggs are stored.

coelom The principal body cavity in most animals, located between the intestinal canal and the body wall.

coprolite Fossilized faeces.

ctenidium (pl. ctenidia) Comb- or feather-like gills, part of the respiratory system of molluscs.

cytotoxicity Meaning toxic to cells.

derived species A species possessing traits or features not found in its earlier ancestors.

detritivore An organism that feeds on dead organic matter.

deuterostome Meaning 'second stomach', this bilaterian group of animals shares an embryonic development plan and includes chordates, echinoderms and hemichordates.

ecdysis The process of moulting and replacing of the cuticle. Moulting is necessary because the cuticle cannot stretch as the animal grows bigger.

ecdysozoa A group of protostome animals that, in addition to the Arthropoda and Nematoda, contains the phyla Tardigrada, Onychophora, Nematomorpha, Priapulida, Kinorrhynca and Loricifera. The name ecdysozoa refers to their distinguishing shared feature, a tough cuticle that is shed as they grow.

endoparasitic An internal parasite, one that lives within the organs and tissues of a host animal.

eusocial/eusociality The most complex form of social behaviour observed in insects such as wasps, ants and bees, which includes cooperative brood care within and across the generations and the division of females into non-reproductive workers and reproductive queens.

flagellum (pl. flagella) A whip-like beating structure, extending from cells, often enabling locomotion.

hydrozoa A class of small, largely marine, invertebrates in the Cnidaria phylum.

kin selection Where workers are related to the offspring and hence increase their evolutionary fitness by helping to rear siblings.

Lophotrochozoa A grouping of bilaterian animal phyla that includes molluscs and annelids.

mesoglea A jelly-like substance without cells found in sponges and cnidarians.

oligochaete A group of annelid worms that have few chaete (body hairs).

phylum (pl. phyla) The first taxonomic division of animals below the kingdom.

polychaete A group of annelid worms that have many chaete (body hairs).

secondary metabolites Organic compounds that are not directly involved in the normal growth, development or reproduction of organisms.

sessile Permanently fixed, immobile.

totipotent Capable of giving rise to any cell type.

1941
Born in Pontiac, Michigan, USA

1963
Graduates from the University of Michigan, Ann Arbor, with a BA in zoology

1964
Receives MS in zoology from Ann Arbor

1967
Awarded PhD in zoology from Ann Arbor

1967–9
Becomes post-doctoral researcher at Harvard University

1967–9
Takes up post of Research Associate at University of Valle, Cali, Colombia

1975–
Becomes Senior Scientist at the Smithsonian Tropical Research Institute, Panama and, from 1979, Costa Rica

1988
Elected member of the US National Academy of Sciences

2003
Publishes *Developmental Plasticity and Evolution*, wins the Hawkins Award for the most outstanding scholarly book of the year

2009
Becomes Vice-chair of the Committee on Human Rights, National Academy of Sciences

2012
Wins Quest Award for Lifetime Achievement, Animal Behavior Society

2018
Publishes paper in *Proceedings of the National Academy of Sciences of the United States of America*, which suggests a link between foetal growth conditions and adult obesity and hence cardiovascular disease

MARY JANE WEST-EBERHARD

Mary Jane West-Eberhard was born into a family who greatly encouraged her curiosity, and through her engagement with a 4-H Club in high school (a US-based network of youth organizations supporting personal development) she was first introduced to entomology. While this did not immediately lead to a deep love of insects, this experience was the key to part-time employment at the Museum of Zoology at the University of Michigan, which ultimately resulted in a career working with insects. West-Eberhard studied zoology at the University of Michigan, Ann Arbor, where she received her MS (1964) and PhD (1967). After two years of post-doctoral work at Harvard University, she started working at the Smithsonian Tropical Research Institute (Panama), living first in Columbia, before moving to Costa Rica, where she has lived since 1979.

West-Eberhard first came to prominence for her work on social wasps, focusing on the evolution of sociality. Wasps are eusocial – that is, they have highly advanced social behaviours, with reproductive queens producing offspring that are cared for by non-reproductive females (workers). She was fascinated by the question of how such behaviour evolved, and argued that non-genetic factors influenced the evolution of eusocial behaviour in addition to accepted explanations arising from kin selection theory.

Her work with social insects resulted in a new focus on phenotypic plasticity – the ability of a genotype to produce different phenotypes (forms). While we commonly think of a genotype producing a single form, this is not always so. Consider aphids. For many species, if predators are around or if there are high levels of crowding, the normally wingless forms can produce winged offspring, which can escape hazards and found new colonies. West-Eberhard argues that such variation can provide grist to the evolutionary mill. In 2003 she published *Developmental Plasticity and Evolution,* which summarized her thoughts on phenotypic variation and how this could affect the evolution of new species. This coincided with developments in molecular biology that resulted in the rapid growth of the field of evolutionary developmental biology; her book pointed to the questions researchers should be asking as they attempt to answer questions of adaptation and speciation using these new 'evo devo' perspectives.

More recently, she has used these perspectives on phenotypic plasticity to explain how poor foetal diet results in increased adult obesity in humans. West-Eberhard argues that what is a beneficial defence mechanism in one situation (infancy), turns out to cause one of the greatest global causes of morbidity and mortality in our species. Zoology is often most controversial, and insightful, when we take lessons learned from nature and use these to turn a lens on our own species.

Mark Fellowes

SPONGES

the 30-second zoology

Being immobile, sponges were at one time thought to be plants, but they are, in fact, animals consisting of relatively few cell types arranged around a skeleton of tough protein called spongin, a particular type of collagen. They possess only two layers of cells separated by a layer of a jelly-like substance called mesohyl. These primitive invertebrates have amazing regenerative properties thanks to totipotent cells called archaeocytes, which are capable of generating all other cell types. Sponges are classed according to the type of spicule (spine) of calcium or silica in their skeletons: Demospongiae, the most diverse, which includes bath sponges; Calcarea, which have calcium spicules; and Hexactinellida, or glass sponges, which have silica spicules. It might seem that sponges would be defenceless when threatened by predators or if competing for space, but they have an arsenal of chemical weapons that have cytotoxic, antibiotic and feeding-deterrent properties. These chemical weapons show striking structural similarities to metabolites of microbial origin, suggesting that microorganisms are the true source of the metabolites or are intricately involved in their biosynthesis. Sponges live in symbiosis with sometimes hundreds of different types of bacteria, and in some species bacteria constitute up to 40 per cent of their biomass.

3-SECOND DISSECTION
Sponges, the sister group of all other animals, are multicellular animals that lack respiratory, nervous or circulatory systems, organs and body symmetry.

3-MINUTE SYNTHESIS
There are approximately 9,000 sponge species worldwide, most of them marine, living attached to solid surfaces or, in some species, the backs of crabs that trade scraps of food for protection. Sponges feed using choanocyte cells with flagella that beat the water to create feeding currents. Their scientific name, Porifera, meaning pore-bearer, refers to the many pores through which sponges absorb food and oxygen and eliminate waste by means of a constant flow of water.

RELATED TOPICS
See also
JELLYFISH, CORALS, ANEMONES & MORE
page 36

ECHINODERMS
page 38

3-SECOND BIOGRAPHY
ROBERT EDMOND GRANT
1793–1874
Scottish anatomist, zoologist and naturalist who discovered that water enters a sponge by small apertures and leaves at certain larger holes; he proved that sponges are animals and is credited with the name Porifera

30-SECOND TEXT
Amanda Callaghan

Sponges are a diverse group of animals but all possess the same feeding system, which is unique to the phylum Porifera.

JELLYFISH, CORALS, ANEMONES & MORE

the 30-second zoology

3-SECOND DISSECTION

Cnidarians are radially symmetrical simple animals distinguished by the presence of specialized stinging cells called cnidocytes, or nematocysts, used for food capture and defence.

3-MINUTE SYNTHESIS

The name Cnidaria derives from the Greek word *knide*, meaning nettle. Cnidarians possess only two cell layers – the ectoderm on the outside and the endoderm that lines the internal cavity that forms the stomach. Between the layers is a cell-less jelly-like substance called mesoglea. All species of cnidarians are aquatic – mostly marine rather than freshwater – and approximately 10,000 species of jellyfish, hydrozoans, corals and sea anemones have so far been described.

The life cycle of Cnidaria – a phylum that includes jellyfish, corals and anemones – is remarkable, with multiple extreme changes in body form. Depending on the class, they develop between sessile polyps, swimming cup-shaped medusae, two kinds of free-swimming larvae and both sexual and asexual reproductive stages. Anthozoan sea anemones and corals lack the medusa stage and exist purely in the polyp form, whereas the Scyphozoan jellyfish and hydra engage in the full multi-stage cycle. Cnidarian bodies are essentially a sac with an opening serving as both mouth and anus surrounded by a ring of tentacles. In jellyfish the tentacles point down; in anemones and corals they point up. Some jellyfish are a metre (39 inches) across with tentacles more than 10 metres (33 feet) long. Arguably the most spectacular cnidarian is the hydrozoan man-o'-war (*Physalia*), a colony of genetically identical polyps and medusae that perform various functions, including that of an inflated sail projecting above the ocean surface to harness the wind. Hard coral polyps secrete a calcium carbonate sheath to house the living animal, and most have a symbiotic relationship with photosynthesizing algae, zooxanthellae, explaining why hard corals are found in shallow waters. Corals are ecosystem engineers, building vast reefs that house other organisms as well as forming a protective buffer between the sea and the coast.

RELATED TOPICS

See also
SPONGES
page 34

ECOSYSTEM ENGINEERS
page 130

3-SECOND BIOGRAPHY
RUTH GATES
1962–2018
Director of the Hawaii Institute of Marine Biology who studied coral bleaching and how, often as a result of warming seas, coral polyps lose their algal symbionts and eventually die

30-SECOND TEXT
Amanda Callaghan

The Cnidaria phylum includes polyps, such as soft-bodied anemones and hard corals, which are fixed to the sea floor, and free-moving medusae, including the many-tentacled jellyfish and man-o'-war.

ECHINODERMS

the 30-second zoology

Echinoderms appear to be

radially symmetrical like a jellyfish, but, while this is true in the adult form, they actually start life as bilateral free-swimming larvae that go on to develop their characteristic radial (often five-sided) symmetry. This embryonic development groups them with the bilaterial deuterostome branch of the tree of life that contains the vertebrates like us. Echinoderms all possess rigid mesodermal plates called ossicles made from calcium carbonate crystal. These take a number of different forms, fusing to make an urchin test, loosely linking in sea stars or forming rings in sea cucumbers. Some species are sessile filter feeders, but others, such as urchins and sea stars, are ferocious mobile predators, sitting on prey and exuding their stomachs and digestive enzymes through their mouths. Soft sea cucumbers, which seem vulnerable to predation, have an arsenal of weapons to fight back. They expel white, sticky, toxic threads called Cuvierian tubules from their anus, which can entangle or even kill a would-be predator. If really annoyed they simply explode their guts through their anus on to the predator and have to wait for them to regenerate. Sea stars also have amazing powers of regeneration and can regrow arms or even split their bodies into two separate animals.

RELATED TOPIC
See also
CHORDATES
page 56

3-SECOND DISSECTION
Echinoderms are the sea urchins, sea stars, brittle stars, sea cucumbers and sea lilies and take their name from the ancient Greek *echinos*, hedgehog, and *derma*, skin.

3-MINUTE SYNTHESIS
Most echinoderms live on the sea bed, although sea stars and sea urchins can often be found in rock pools on the sea shore. A unique feature of echinoderms is their water vascular system. This is a complex series of channels that move seawater through a pore called the madreporite to generate differential water pressures. This moves their hollow tube feet, pumping them up and down like a thin balloon.

3-SECOND BIOGRAPHIES
GEORGES CUVIER
1769–1832
French naturalist and zoologist who established the scientific fields of palaeontology and comparative anatomy; he was the first to describe the sea cucumber behaviour of expelling toxic threads

ERNST HAECKEL
1834–1919
German naturalist, zoologist, physician and philosopher; he was an extraordinary artist who discovered and named thousands of new species

30-SECOND TEXT
Amanda Callaghan

The many-armed echinoderms are all marine invertebrates in possession of a spiny shell or leathery skin.

FLATWORMS

the 30-second zoology

3-SECOND DISSECTION
Platyhelminthes are bilaterian flattened worms that range in size from microscopic to several metres in length; many are parasites while others inhabit marine, fresh water or damp terrestrial habitats.

3-MINUTE SYNTHESIS
Placing Platyhelminthes on the tree of life has been controversial and hindered by the absence of a fossil record, although fossilized eggs were discovered in a 270-million-year-old shark coprolite. Modern analyses show that Platyhelminthes, molluscs and annelids share a common ancestor. Platyhelminthes may have evolved from a more complex ancestor, losing such features as a through-gut to adapt to an endoparasitic lifestyle. The few Platyhelminthes with a gut have no anus; they eat and excrete via their mouths.

Platyhelminthes – or flatworms – are soft-bodied, dorsoventrally flattened worms with no body segmentation. They lack a through-gut, coelom, circulatory and respiratory systems and are able to absorb oxygen and nutrients by direct diffusion through their skin – handy if you live inside other animals as most Platyhelminthes do (including such well-known species as the highly specialized tapeworms that live in vertebrate guts). Adults are little more than gonads with a head full of hooks with which they attach themselves to and feed from their host's intestines. Trematodes, or flukes, are also parasites of vertebrates. The life cycle of most flukes involves being hosted by snails and vertebrates. Larval forms released from infested snails into fresh water infect people and animals, either by direct contact with water or by eating undercooked fish, crustaceans or poorly washed raw watercress. Not all Platyhelminthes are parasitic. The Turbellaria includes non-parasitic, or free-living, flatworms called planarians. These worms have a remarkable ability to regenerate body parts by virtue of having retained specialized stem cells into adulthood. This is great for the worm but also for scientists in need of numerous flatworms for research. Instead of breeding planarians they can simply cut these worms up to generate a population to work with.

RELATED TOPICS
See also
SEGMENTED WORMS
page 44

MOLLUSCS
page 48

3-SECOND BIOGRAPHY
KARL GEGENBAUR
1826–1903
German comparative anatomist and strong supporter of the theory of evolution who first coined the name Platyhelminthes and divided the phylum into four classes

30-SECOND TEXT
Amanda Callaghan

Many flatworms, such as tapeworms and flukes, are happy to inhabit the bodies of other organisms – from humans to snails.

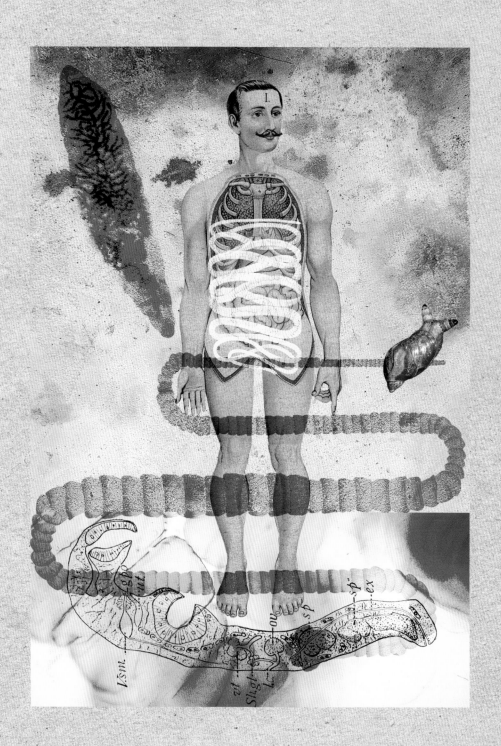

ROUNDWORMS

the 30-second zoology

Although the common name for nematodes is roundworms, 'worm' is a vernacular term related to their body shape and is not an indicator of evolutionary relationships. Roundworms are on the same branch of the tree of life as arthropods in a clade called Ecdysozoa. Like insects and crustaceans, they shed their cuticle to grow by a process called ecdysis. Unlike arthropods, they lack the polysaccharide chitin in their skin and instead have the animal protein collagen, also found in human skin. Nematodes have a well-developed body cavity with a complete digestive tube but lack respiratory or circulatory systems. They move with a characteristic wiggling S shape because they have muscles that run only lengthways along their bodies. Nematode heads can have mouths with hooks, jaws or spines and chemoreceptor sense organs. Many species are parasitic and are significant agricultural pests of plants and veterinary and medical pests of animals – anyone owning a cat or dog will have treated their pets with drugs to kill intestinal roundworm parasites. Humans are also susceptible to roundworm infestation, and filarial roundworms, transmitted by mosquitoes, can infect lymph nodes, passing on serious debilitating diseases such as elephantiasis, which can cause legs and other extremities to swell to enormous proportions.

3-SECOND DISSECTION
Roundworms are unsegmented worms, round in cross-section but tapered either end, that occur in almost every habitat, including inside plants and other animals.

3-MINUTE SYNTHESIS
Nematodes are one of the most abundant animals on the planet and are of great economic importance, yet few people will ever have seen one. Most nematodes are less than one millimetre in length, but they are essential for soil aeration and the recycling of organic and mineral matter. Although only approximately 23,000 species have been formally described, it is estimated that there are up to a million nematode species on the planet.

RELATED TOPICS
See also
ARTHROPODS
page 46

SEGMENTED WORMS
page 44

3-SECOND BIOGRAPHY
NATHAN AUGUSTUS COBB
1859–1932
American researcher, known as the father of nematology in the USA, who identified more than 1,000 species of nematode. He famously commented that if everything apart from nematodes was magically removed from the planet 'our world would still be dimly recognizable ... we should find its mountains, hills, vales, rivers, lakes, and oceans represented by a film of nematodes'.

30-SECOND TEXT
Amanda Callaghan

Roundworms can live, feed and reproduce in the gut without any apparent symptoms in the host, but where they infest the lymph nodes, serious diseases can develop.

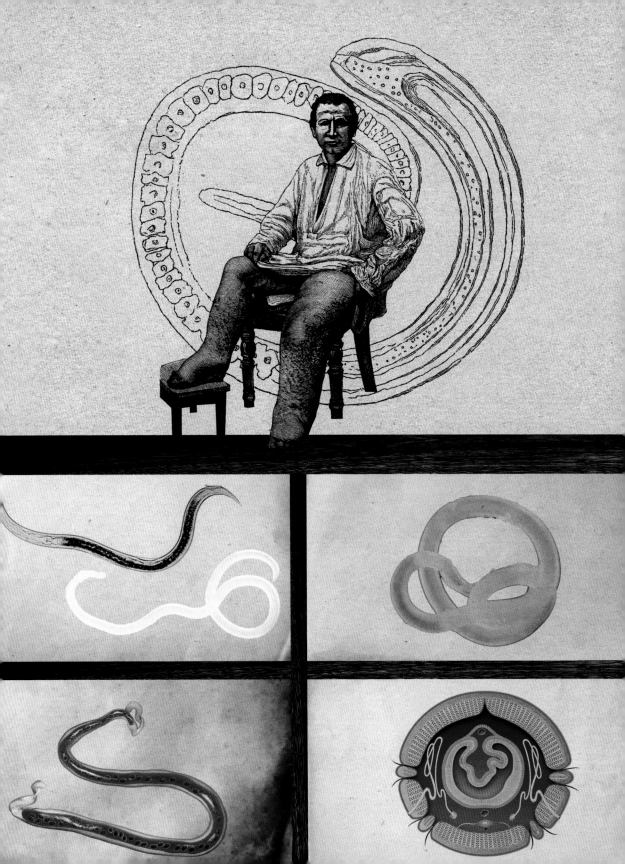

SEGMENTED WORMS

the 30-second zoology

Annelida – or segmented worms

– are an ecologically diverse phylum of animals within the clade Lophotrochozoa. One annelid group, the Clitellata – named for the mucus-covered clitellum that provides a cocoon for their eggs – includes blood-sucking leeches and earthworms. Leeches have lost many annelid features and most use suckers to latch on to prey, cutting into the flesh with sharp jaws or sucking through a proboscis. Oligochaete earthworms burrow through soil feeding on decomposing organic matter. Polychaetes, or bristle worms, are by far the most abundant annelid group. These diverse and colourful marine annelids range from one millimetre to three metres (3⁄64 inch–10 feet) in length and include burrowing, crawling and sessile species as well as spectacular predators. Their segments have little paddles called parapodia, which are used for swimming or walking. One, the ferocious bobbit worm, can grow up to a metre in length and lies buried in the seabed with only its sharp open jaws poking above the sand. Bobbit worms have the strength and speed to grab a passing fish and pull it down into its burrow. *Glycera polychaetes* – bloodworms – are popular bait for fishermen. Their bite contains a neurotoxin that causes uncontrollable muscle twitching, effectively paralysing prey. This wouldn't kill a human but can be very painful.

RELATED TOPICS

See also
ARTHROPODS
page 46

MOLLUSCS
page 48

3-SECOND DISSECTION
Annelid worms are characterized by the segmentation of their cylindrical bodies, which, from the Latin word *annulus*, meaning ring, gives the phylum its name.

3-MINUTE SYNTHESIS
Because of their diversity, annelids are found in a huge variety of moist habitats, including marine, freshwater and terrestrial environments. Some species are suspension feeders, others are detritivores, deposit feeders, scavengers, herbivores or carnivores. With at least 17,000 described species, annelids are so morphologically diverse that there is no one feature that defines them all. However, most annelids have segmentation, bilateral symmetry, chitinous bristles called chaetae and a hydrostatic skeleton.

3-SECOND BIOGRAPHY
KATHARINE BUSH
1855–1937
The first woman at Yale to receive a PhD in the sciences; her thesis was on the sabellid and serpulid polychaetes collected by the Harriman Alaska Expedition of 1899, which, as a woman of her time, she had been unable to join directly

30-SECOND TEXT
Amanda Callaghan

From the fish-snatching bobbit worm to the common earthworm, the annelids populate both marine and terrestrial habitats.

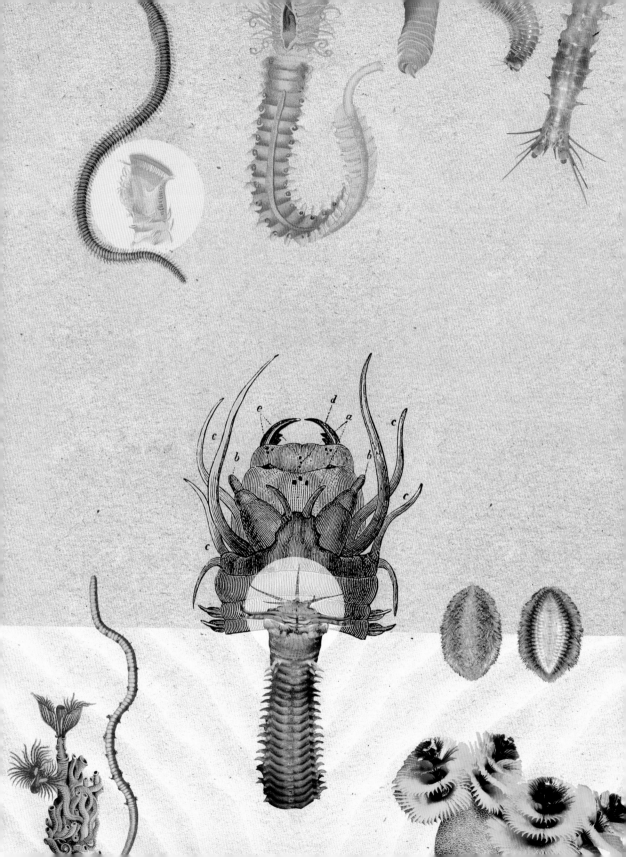

ARTHROPODS

the 30-second zoology

Arthropods first evolved in the ancient seas around 600 million years ago, and yet their most successful descendants are the terrestrial insects. Insects, which evolved from crustacean ancestors, have three pairs of legs, a head, thorax and abdomen, and most species in the 27 insect orders have wings. They have filled almost every conceivable non-marine niche, adapting to both endoparasitic and ectoparasitic lifestyles and exploiting multiple food sources including plants and vertebrate blood. Public perception of insects is often negative, yet only around 3 per cent of insect species are pests and many, such as pollinators, are extremely beneficial. Nearly all crustaceans, including lobsters, crabs and shrimps, live in salt or fresh waters; woodlice are the only truly terrestrial crustacean. Crustaceans have exoskeletons hardened with calcium carbonate and a highly varied body plan, ranging from barnacles that cement themselves on to rocks to enormous marine crabs. Myriapods, meaning many legged, are the cylindrical millipedes and flattened predatory centipedes. Chelicerates are the terrestrial arachnids (spiders, scorpions and ticks) and the marine horseshoe crabs and sea spiders. Most chelicerates have only two body parts – the cephalothorax, a fusion of the head and thorax, and the abdomen – and at least four pairs of legs.

RELATED TOPICS
See also
ROUNDWORMS
page 42

3-SECOND DISSECTION
Approximately 80 per cent of all animal species are arthropods, which makes them the most successful animals on the planet.

3-MINUTE SYNTHESIS
Arthropods (from ancient Greek *arthron*, jointed, and *podes*, feet) – which are grouped into the insects, chelicerates, crustaceans and myriapods – are a hugely successful phylum of sophisticated animals with well-developed respiratory systems, sensory organs, a true body cavity (coelom) and a protective and versatile exoskeleton made of chitin. As members of the Ecdysozoa clade they grow by shedding their cuticle. They are extremely adaptable and are found in all environments and at extremes of temperature, pressure and salinity.

3-SECOND BIOGRAPHY
MIRIAM ROTHSCHILD
1908–2005
Known as the 'Queen of the Fleas', Miriam Rothschild FRS, was a self-taught world authority on fleas, bees and butterflies. She discovered that the life cycle of the rabbit flea, vector of myxomatosis, was linked to the host's sex hormones. Famously she kept live fleas in plastic bags in her bedroom so that she could 'see what they are doing and so children do not annoy them'.

30-SECOND TEXT
Amanda Callaghan

Arthropods are the most diverse of all animals on Earth. They include herbivores, predators and detrivores and occupy terrestrial, aerial and aquatic habitats.

MOLLUSCS

the 30-second zoology

More than 80 per cent of all mollusc species are gastropods, the snails and slugs, which have a large foot that makes up most of the visible soft flesh. Their shells tend to be coiled or spiralled, and the body can be withdrawn into it, although in some the shell has been significantly reduced or even, as with slugs, lost completely. The 180-degree rotation or twisting of their bodies during development, called torsion, is a gastropod characteristic resulting in an anus that is located above the mouth. Only molluscs have a radula, a specialized toothed structure that is used to rasp food from a hard substrate or bore into prey. Cephalopods, such as the octopuses, have both a radula and beak-like jaws, while cone snails have adapted their radula into toxin-soaked harpoons that they can shoot at prey or predators. Bivalves lack a radula and have instead adapted their ctenidia into a net-like mesh to filter-feed from the water. Cephalopods are intelligent marine predators that are highly derived, with prehensile tentacles, ink sacs and a muscular funnel called a siphon that expels water for rapid movement. By contrast, many bivalves, whose principal feature is their two-hinged shells, burrow into sediment or attach themselves to objects on the seabed.

3-SECOND DISSECTION
Molluscs form a
very diverse group of
mostly marine animals
characterized by having a
mantle that secretes
calcium carbonate for shell
formation, a radula for
feeding and gills (ctenidia).

3-MINUTE SYNTHESIS
The molluscs (Mollusca)
form the second largest
animal phylum after the
arthropods, comprising
around 80,000 described
species grouped into eight
extant classes. Most
people are familiar with the
bivalves, snails and slugs,
octopuses and squid but
have probably not come
across the other classes,
which include worm-like
forms and chitons covered
in articulated plates. Some
snail and bivalve species
live in fresh water, but only
the snails and slugs have
managed to survive on
dry land.

RELATED TOPIC
See also
SEGMENTED WORMS
page 44

3-SECOND BIOGRAPHIES
JEANNE VILLEPREUX-POWER
1794–1871
French amateur naturalist
interested in cephalopods; she
preferred to study live animals
and is credited with the
invention of the aquarium

GEORG EBERHARD RUMPHIUS
1627–1702
German-born botanist who was
the first to publish a mollusc
taxonomy and gave names to
groups such as the gastropods
and bivalves

THOMAS SAY
1787–1834
American naturalist, known
as the father of American
conchology, who collected,
studied and described many
insects, molluscs and reptiles

30-SECOND TEXT
Amanda Callaghan

*Molluscs are soft-
bodied invertebrates,
some of which
are encased in
a protective shell.*

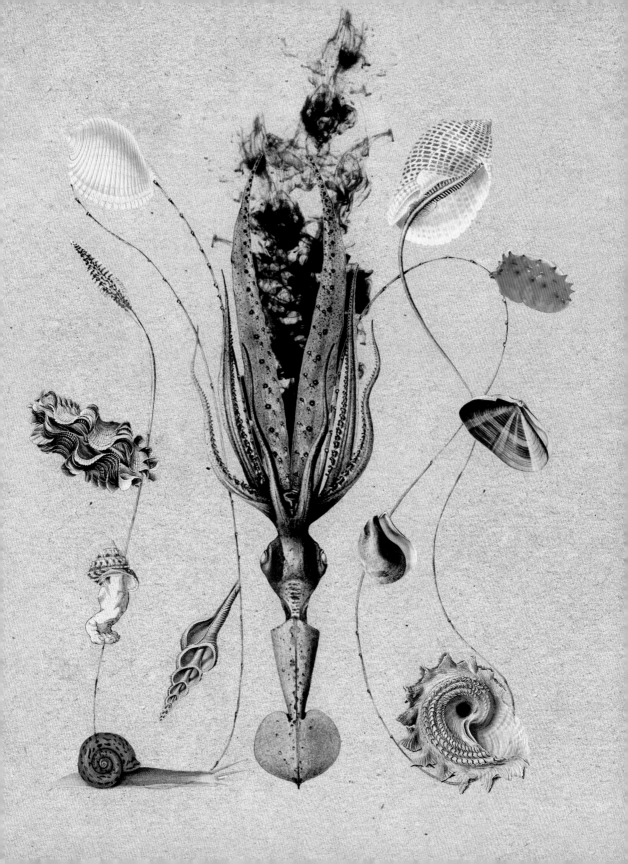

THE VERTEBRATES

THE VERTEBRATES
GLOSSARY

amniotes Term for the tetrapod vertebrates that either lay eggs on land (birds and reptiles) or retain their fertilized eggs in the mother's body (mammals), which distinguishes them from egg-laying amphibians and fish.

anadromous Term describing fish, such as salmon, that leave ocean environments to spawn in rivers and lakes.

binomial nomenclature The classification system whereby organisms are named using a two-word identifier, the first being genus and the second the species.

bioindicator A species or group of species that indicate changes in the environment, often by fluctuations in population number.

caecilians Secretive, limbless tropical amphibians related to salamanders.

cephalochordates Small, segmented invertebrate marine animals that possess a notochord (see below).

endemic Plants and animals found in a place. The term endemic can apply to a specific or a broad region.

endostyle A structure found in lower chordates and larval lampreys that aids in filter-feeding.

Hox gene A set of genes that codes for the body plans of organisms.

herpetologist One who studies amphibians and reptiles.

ichthyology The branch of vertebrate zoology concerned with the study of fishes.

monotremes Odd, egg-laying mammals, such as the platypus and echidna.

notochord A stiff, dorsal rod running the body length of all embryonic vertebrates and some adult invertebrate animals.

photobacteria A group of mostly marine bacteria that are bioluminescent (meaning they have the ability to glow in the dark).

placoid scale The small, hardened scales made of enamel found on the bodies of sharks, skates and rays.

synapsids Ancient reptiles, the ancestors of mammals, having one opening in the skull.

tetrapodomorph Any four-limbed vertebrate. The Tetrapodomorpha is a clade that includes all four-limbed vertebrates and their nearest sarcopterygian relatives (lobe-finned fish), of which the coelacanth is the only extant (living) species.

tetrapods Vertebrates with two pairs of limbs. Tetrapods include all existing and extinct amphibians, reptiles (including dinosaurs and birds) and mammals.

tuatara Lizard-like reptiles that are endemic to New Zealand.

tunicate A small, marine invertebrate chordate, which feeds by siphoning water in and out of its body.

urochordate A synonym for tunicate (see above).

4 May 1922
Born in New York City, USA

1942
Graduates with Bachelor's degree in zoology from Hunter College, New York

1946/1950
Receives MA and PhD from New York University; while undertaking her studies she conducted research at the Scripps Institution of Oceanography in La Jolla, California, the American Museum of Natural History, New York, Woods Hole Marine Biological Laboratory, Massachusetts, and also at the Lerner Marine Laboratory in Bimini, Bahamas

1949
Studies fish populations in Guam, the Marshall Islands, the Palau Islands, the Northern Mariana Islands and the Caroline Islands

1953
First book, *Lady with a Spear*, is published; it recounts her experiences in the Pacific

1966/1968
Teaches at the City University of New York and University of Maryland, College Park, Maryland, USA

1969
Publishes *The Lady and the Sharks* about her work in shark biology and conservation

1999
Retires from teaching but continues to teach one zoology course per semester for many years

2000
Becomes Senior Scientist, Director Emerita and Trustee at Mote Marine Laboratory, Sarasota, Florida, USA

25 February 2015
Dies at the age of 92 in Sarasota, Florida, USA

EUGENIE CLARK

Eugenie Clark seemed destined to be a marine biologist. During her grade- and high-school days in New York she often wrote reports about marine life for her assignment topics. She visited the aquarium every Saturday and was mesmerized by the life and careers of famous marine scientists and explorers. She decided early on that her life goal was to become one.

She applied to Columbia University for postgraduate study, but, being married with small children, her application was rejected as it was assumed she would leave to focus on family. Instead, she earned her MA and PhD degrees from New York University.

Under an Office of Naval Research programme to undertake scientific research in Micronesia in 1949, Clark studied fish populations in the Pacific. She also conducted ichthyological studies at the Marine Biological Station in Hurghada, Egypt, where she discovered that one fish, the Moses sole, exudes a natural shark repellent. Her first book, *Lady with a Spear* (1953) – which recounted her experiences while conducting this work – met with acclaim and inspired financial backers to fund her research, and she remained a prolific writer throughout her career, authoring some 175 scientific articles.

Among her discoveries was the fact that, contrary to what was believed at the time, sharks need not move to breathe oxygen – although this myth persists today. She conducted experiments where she was able to train sharks to perform simple tasks, indicating that they possessed intelligence and were not just simple eating machines. Concerned that sharks were being unfairly persecuted because of their reputation, Clark gave public lectures on shark behaviour as well as working to develop many television specials on marine conservation. She also made over 70 dives in a submersible, a little-known technology at the time. Of her many accomplishments, developing scuba equipment as a research tool is significant, and she received numerous awards and honours, including three honorary degrees, in recognition of her many contributions to marine biology and conservation.

Perhaps the greatest distinction for a zoologist is to have a new species named in recognition of their work, and Eugenie Clark, popularly known as 'the Shark Lady', had many, including *Callogobius clarki*, *Sticharium clarkae*, *Enneapterygius clarkae*, *Atrobucca geniae* and, notably, *Squalus clarkae* – also known as 'Genie's dogfish'.

Peter Capainolo

CHORDATES

the 30-second zoology

3-SECOND DISSECTION
Chordates are a diverse phylum of animals that vary from mat-like sea squirts on the undersides of rocks to dynamic and (usually) intelligent humans.

3-MINUTE SYNTHESIS
Chordates are grouped with the Hemichordata and Echinodermata phyla. Hemichordates are worm-like marine animals that are split into Enteropneusta (acorn worms) and Pterobranchia (colonial sea angels). They share some features with Chordates, which are divided into three groups: the urochordate tunicates, cephalochordates and vertebrates. Primitive chordates have an endostyle, an organ at the base of the pharynx that secretes mucus used for feeding. In vertebrates this has disappeared and been repurposed into the thyroid gland.

Chordate bodies vary widely in form, but all are stiffened by a close-packed cylinder of cells within a toughened membrane called a notochord. In vertebrates the notochord is replaced by the vertebral column, or spine. Chordates have a dorsal nerve cord and, at some stage in their lives, most have longitudinal muscles along the length of the body. These move against the stiffened notochord or vertebrae, allowing animals to bend and become highly active. If you were to draw a basic chordate, it would probably resemble the tiny fish-like marine animal called a cephalochordate, also known as a lancelet. These are important animals to developmental biologists who have studied their Hox genes as an example of an ancestral form. However, cephalochordates are not *our* ancestors. Perhaps surprisingly, molecular analysis has revealed that the urochordates, also known as sea squirts, are more closely related to vertebrates. These sac-like animals appear to be simple creatures that filter feed through two main siphons. However, it is in their free-swimming larval form that their heritage becomes obvious, with a notochord, nerve cord, pharynx and post-anal tail. The remaining Vertebrata group is known as Craniata, since all have a bony or cartilaginous brain case but not all have vertebrae.

RELATED TOPICS
See also
GENES
page 16

ECHINODERMS
page 38

FISH
page 56

3-SECOND BIOGRAPHY
PETER HOLLAND
1963–
British evolutionary biologist and zoologist whose research focuses on how the evolution of animal diversity can be explained through the evolution of the genome

30-SECOND TEXT
Amanda Callaghan

Primitive chordates, such as sea squirts and lancelets, mark the transition between invertebrates and vertebrates. The lamprey is a living example of an ancient lineage of jawless fish that preceded true fish.

FISH

the 30-second zoology

Spanning 60 million years, the Devonian is known as the 'Age of Fishes' because of the wide diversity of major fish groups found in fossils from the period. Sharks and rays of all sizes – which have skin covered by placoid scales and a skeleton composed of cartilage rather than ossified bone – were abundant, and their descendants still hunt in the world's oceans today. Also first appearing in the Devonian oceans were the bony fish, or Osteichthyes. As the name suggests – and unlike sharks and rays – the skeletons of bony fish are made of bone, and their scales are of variable shape and composition depending on the species. Most of the approximately 28,000 species of fish currently known to science are Osteichthyes, and they are divided into two classes: the Sarcopterygii, or lobe-finned fish, and the Actinopterygii, or ray-finned fish. The Sarcopterygii are important evolutionarily, as they appear to be ancestral to a group of vertebrates that made the transition from water to land c. 390–360 million years ago. Actinopterygii are by far the most numerous, comprising around 99 per cent of all fish species. All these animals exchange oxygen and carbon dioxide gases using gills, and many possess a swim bladder, which is important in maintaining buoyancy in water.

RELATED TOPICS
See also
EUGENIE CLARKE
page 54

AMPHIBIANS
page 58

3-SECOND DISSECTION
Fish display a remarkable diversity and represent an important evolutionary transition of vertebrates from aquatic to terrestrial environments that occurred c. 390–360 million years ago.

3-MINUTE SYNTHESIS
Ichthyologists have no shortage of fascinating phenomena to study. Catfish and sharks possess organs that detect electric current emanating from aquatic life forms. South American knifefish produce electric currents for navigation and social interaction. Several species of salmon are anadromous, sometimes migrating huge distances from sea to fresh water in order to spawn. Deep-sea angler fish possess a glowing bioluminescent 'lure' on their heads to attract prey; the glow is produced by photobacteria.

3-SECOND BIOGRAPHIES
PETER ARTEDI
1705–1735
Swedish naturalist, known as the father of ichthyology; a colleague of Carl Linnaeus who formalized the system of binomial nomenclature

MARJORIE EILEEN DORIS COURTENAY-LATIMER
1907–2004
South African museum curator who discovered a specimen of an extant Sarcopterygian fish, the coelacanth, in 1938

30-SECOND TEXT
Peter Capainolo

Fish are finned, cold-blooded animals that live entirely in salt or fresh water and breathe through gills.

AMPHIBIANS

the 30-second zoology

3-SECOND DISSECTION
Amphibians evolved from
fish that could crawl as well
as swim, giving rise to the
first terrestrial creatures
with four limbs – the
precursors of reptiles.

3-MINUTE SYNTHESIS
Amphibians are sensitive
to changes in the
environment. Since they
breathe through their skin
they absorb small particles
of possibly harmful
substances, and ecologists
have observed that subtle
changes in the environment
are manifested by a
decrease in frog and other
amphibian populations.
Occasionally poisons or
infectious agents cause
serious mutations in frogs,
including deformities and
missing or extra limbs.
These phenomena have
led biologists to consider
these animals to be
bioindicator species.

Some 370 million years ago, four-limbed fish began to make short forays on to the land, giving rise to the first amphibians, cold-blooded animals able to breathe both in and out of water. Eventually their descendants became quasi-terrestrial, feeding on land but returning to water to lay eggs. These eggs were non-amniotic, like those of fish – lacking a hard shell and producing aquatic larvae that gradually morphed into the adult form – and modern amphibians still retain the physical characteristics and reproductive strategies of their ancestors. Today herpetologists recognize some 7,000 species of amphibians, most of which are frogs, and there is great diversity in size, colour and survival strategies within the three amphibian groups: frogs, salamanders (including newts) and caecilians. All amphibians (albeit toads to a lesser extent) possess glands in their skin which secrete a smooth, slimy layer of mucus that keeps the skin from drying out. Most have lungs but also absorb oxygen through their skin, while some aquatic species, and the larval stages of all, use gills for gaseous exchange. Many adult amphibians feed on insect life, but some of the larger toads and salamanders will catch shrimp, crabs and even mice and birds. The least familiar group are the limbless, snake-like caecilians, which are reclusive animals that live in soil in tropical regions.

RELATED TOPIC
See also
FISH
page 58

3-SECOND BIOGRAPHIES
MARÍA CRISTINA ARDILA-ROBAYO
1947–2017
Colombian herpetologist
who described 28 new species
of amphibians from Colombia
and has four species named
after her

DAVID KIZIRIAN
1960–
American herpetologist who
has discovered and described
many new species of
amphibians and reptiles

30-SECOND TEXT
Peter Capainolo

Most amphibians live both in water and on land, but during their egg and larval stages they are wholly aquatic before becoming terrestrial adults that breathe with lungs.

REPTILES

the 30-second zoology

Herpetologists recognize

approximately 10,000 species of reptiles, the vertebrate class that includes lizards, snakes, turtles, tortoises, crocodiles, alligators and tuatara. Reptiles evolved about 310 million years ago and were able to flourish on land because of several important physical adaptations that their amphibian antecedents lacked. Although still cold blooded, reptiles developed functional lungs and skin with a leathery or scaly surface. Some extinct forms, such as ichthyosaurs and plesiosaurs, returned to the oceans and developed remarkably fish-like bodies, and several species of terrestrial dinosaurs grew to immense size, possibly influenced by diet, climate and other factors. Reptiles were the first group of tetrapods to lay amniotic eggs with an external protective shell, freeing these animals from the need to lay eggs in water. This led to reptiles radiating into all major ecological niches and further evolving into myriad diverse species. Like other vertebrate groups the majority of reptile species are found in tropical and subtropical regions; those occupying colder environments will spend much of the winter months in a torpid state under leaves and mud. Reptiles in this state have very low heart and breathing rates, an evolutionary strategy that conserves energy for breeding activities during the warm spring and summer months.

3-SECOND DISSECTION
Reptiles are cold-blooded, scaly-skinned vertebrates that move by crawling or wriggling and that lay soft-shelled eggs on land.

3-MINUTE SYNTHESIS
Several reptile subgroups appear little changed from their distant ancestors, and they display specialisms that have ensured their survival. Crocodiles have adaptive metabolisms that enable them to go without food for months. Snakes hunt by flicking their tongues to detect particles shed by prey animals, which are then dispatched by constriction or a venomous bite and swallowed whole. The huge Komodo dragon of Indonesia has venom glands in its jaws to kill prey as large as pigs and small deer.

RELATED TOPICS
See also
PHYLOGENETICS & DIVERSITY
page 22

AMPHIBIANS
page 60

3-SECOND BIOGRAPHIES
WILLIAM DOUGLAS BURDEN
1898–1978
American naturalist and filmmaker who shot the first footage of Komodo dragons and collected the first specimens

LESLIE JANE RISSLER
1969–
American herpetologist and writer who focuses on outreach work with the general public about evolution

30-SECOND TEXT
Peter Capainolo

Important evolutionary developments that are features of reptiles, notably an egg with a watery environment housed in a shell, paved the way for a fully terrestrial existence.

BIRDS

the 30-second zoology

The vertebrate class Aves reveals some of the greatest diversity of any class. Approximately 10,000 species are living today, ranging in size from the 2.1-metre (almost 7-foot) omnivorous ostrich to the 6-centimetre (2½-inch) nectar-feeding bee hummingbird. Birds evolved during the Jurassic period (201–145 million years ago), close relatives of maniraptoran dinosaurs such as the velociraptor. Modern birds are warm blooded, have feathers that grow out of the skin, and lay hard-shelled amniotic eggs. They occupy all ecological niches. Most species can fly, and physical adaptations that allow life on the wing include the absence of teeth and urinary bladder, a reduced, fused skeleton and hollow bones. Most are also sedentary and reside in the tropics, but some that breed in northern latitudes make arduous migrations south every autumn to overwinter in warmer climes. Many species possess high levels of intelligence, using song, display and mimicry to attract a mate, building intricate nests and navigating over land and sea using Earth's magnetic field. The beaks and scale-covered legs and toes of birds are adapted to food preference, habitat and locomotion: meat-eating raptors have sharp curved bills for tearing flesh and toes tipped with sharp talons for gripping prey; seed eaters have short, strong conical beaks; woodpeckers use their chisel-like bills and long sticky tongues for extracting insects while clinging to trees.

3-SECOND DISSECTION
With the ability to fly, birds rapidly radiated throughout the world, monopolizing food sources and breeding ranges to become an enormously successful class of vertebrates.

3-MINUTE SYNTHESIS
The remarkable variation among birds makes them favourite subjects for scientific study. There are birds that can dive and swim, ones that burrow but no longer fly and a few, such as the swift and the albatross, that spend almost their entire lives on the wing. One of the strangest birds is the flightless kiwi, which is endemic to New Zealand. It is very mammal-like; its feathers resemble hair, and it possesses a keen sense of smell, a trait absent in most birds.

RELATED TOPICS
See also
PHYLOGENETICS & DIVERSITY
page 22

ISLAND ODDITIES
page 26

3-SECOND BIOGRAPHIES
JOHN GOULD
1804–81
British naturalist, ornithologist and taxidermist who studied Darwin's finches

ELLIOTT COUES
1842–99
American physician and ornithologist, known for major scientific writings on birds

MARGARET MORSE NICE
1883–1974
Revered American amateur ornithologist who made seminal contributions to the field

30-SECOND TEXT
Peter Capainolo

Birds have adapted to most ecological niches and their ability to fly, walk, climb, run and swim sets them apart from other classes.

MAMMALS

the 30-second zoology

Ancient reptiles known as synapsids are thought to have given rise to the vertebrate lineage that evolved into modern mammals. All mammals are warm blooded and have hair or fur; all are tetrapods and most are placental, meaning their young develop in the uterus before birth. Humans are placental mammals, as are whales, illustrating how diverse mammals are but also how closely related, based on a similar reproductive strategy. Once born, infant mammals remain highly dependent on their mothers and need to be fed on their mother's milk. Marsupials, the group of mammals that includes kangaroos, give birth to undeveloped foetus-like young that make their way into the mother's abdominal pouch. There they feed from milk glands until they are old enough to emerge and fend for themselves. A third group of mammals, the monotremes – which includes the platypus and spiny anteater of Australia and New Guinea – are the only mammals to be born from eggs laid outside the mother's body. Milk is provided to their young through pores in the skin, as monotreme milk glands are relatively small compared with other mammals, and mothers have no nipples. It is this provision of milk and the associated high level of parental care that truly defines what it is to be a mammal.

3-SECOND DISSECTION
Reproductive strategies in the mammalian subclasses reveal an evolutionary path: egg-laying in monotremes is primitive; semi-foetal births in marsupials intermediate; and placental development most advanced, but all mammalian mothers produce milk.

3-MINUTE SYNTHESIS
The largest living mammal is the blue whale and among the tiniest are smaller-than-thumb-sized species of shrew. Whales and shrews may be hugely divergent in appearance and behaviour, but under the skin are similarities. Even marine-dwelling mammals feed their young milk: a blue whale calf daily drinks about 350 litres (77 gallons) of milk in its first few months. Milk and parental care has allowed mammals to adapt to some of the most extreme habitats on Earth.

RELATED TOPICS
See also
NATURAL SELECTION
page 18

PHYLOGENETICS & DIVERSITY
page 22

3-SECOND BIOGRAPHIES
JOSEPH GRINNELL
1877–1939
The first director of the Museum of Vertebrate Zoology, University of California, Berkeley, who conducted major surveys of mammal populations

DANIELLE 'HOPI' ELIZABETH HOEKSTRA
1972–
American evolutionary biologist who studies the genetic basis of adaptation using wild rodent populations

30-SECOND TEXT
Peter Capainolo

Marsupials, mammals and monotremes have more in common than meets the eye.

PHYSIOLOGY

abiogenesis The origin of life from non-living matter, the transition from chemical processes to biological ones.

cones and rods Photoreceptor cells in the eye. See opsin.

cytotoxins Substances having a toxic effect on cells, leading to cell and organ death, a condition known as necrosis.

ectoderm The outer layer of tissue in an embryo that develops into the skin and nervous system.

endoderm The innermost layer of tissue in an early embryo which will eventually form the gut.

gastrulation The movement of tissue in a single (tissue) layered blastula stage embryo to form a multi (tissue) layered gastrula.

germ layers The ecto-, meso- and endoderm tissues that form specific tissues and organs throughout development.

haemotoxins Toxins that damage red blood cells; can often alter how blood clots and cause organ damage and eventual failure. This can be either through the toxin preventing of clotting, leading to internal bleeding, or by causing blood cells to clump together at much higher rates than normal, blocking blood vessels.

imaginal discs A patch of tissues, that form in a larval insect, which develop into adult structures, such as wings, antennae and legs, at metamorphosis.

mesoderm The middle layer of the tissues in an embryo, develops into bone, cartilage and muscle.

neurotoxins Poisons that act on the nervous system, disrupting the normal chemical signals sent between nerve cells or along the neuron itself.

organizer A region of a developing embryo that secretes signalling molecules that direct and 'organize' development.

worganogenesis The formation of organs, from defined tissues from specific germ layers.

opsin A light-sensitive protein found in the photoreceptor cells (cones and rods) of the eye. Different opsins are sensitive to different wavelengths (energies) of light and combinations of different opsins allow us to see different colours.

polarization In the optical sense, the direction in which a wave of light oscillates while moving. The term also defines the spatial biological specificities in a cell during development, providing the tail–head axis, for example (see page 16).

resorption Refers to the reabsorption of larval tissue (such as the tail and gills), as the juvenile develops.

3 August 1915
Born in Southampton,
New York, USA

1934–42
Studies at Harvard
University,
Massachusetts, USA

1941
Developed the concept of
echolocation with Robert
Galambos

1942
Asked to contribute to
developing a bat bomb by
the US National Defense
Research Committee

1946–53
Teaches zoology at
Cornell University, Ithaca,
New York, USA

1952
Elected a fellow of the
American Academy of
Arts and Sciences

1953–65
Returns to Harvard
University as professor of
zoology

1958
Awarded the Daniel
Giraud Elliot Medal by the
US National Academy of
Sciences for meritorious
work in zoology

1965
Joins Rockefeller
University, New York City,
USA

1976
Publishes *The Question
of Animal Awareness*,
arguing that animals are
conscious beings

1986
Retires from Rockefeller
University

7 November 2003
Dies at the age of 88
in Lexington,
Massachusetts, USA

DONALD GRIFFIN

Few scientists can claim to have changed the way that people view animals, but Griffin did it twice: first, when his studies revealed the way some animals navigate by using echolocation and, second, with his work in the field of cognitive ethology.

Donald Redfield Griffin was born in Southampton, New York, in 1915. As a youth he was a keen naturalist, and he first worked with bats while still in high school, researching the migration patterns, homing instincts and lifespan of brown bats, *Myotis lucifugus.*

His academic career started with research in comparative physiology. While an undergraduate at Harvard he worked with fellow student Robert Galambos to begin to unravel the secrets of how bats navigated in the dark, using a small dark room in which obstacles could be placed. Realizing that bats could use reflected sound to identify and detect objects, Griffin described how this worked and gave this previously unknown way of 'seeing' the world a name: echolocation. It was a contentious idea in scientific circles, but the discovery was a major breakthrough – importantly, it also informed the development of radar and sonar.

One surprising development of his work with bats came in 1942 when he was asked by the US National Defense Research Committee to evaluate an idea for a 'bat bomb'. This called for tiny time-delayed explosive devices to be attached to bats that would be dropped by aeroplanes at night over Japanese cities. Seeking shelter during the day, the bats would roost in buildings, and the explosives would detonate and set off fires. Griffin helped develop the proposal and even collected Mexican free-tailed bats to test their load-bearing capacity. Bat bombs were never used in combat, however, and he later questioned the ethics of using animals in such a way.

Griffin pioneered cognitive ethology, a field concerned with the influence of conscious awareness and intention in animal behaviour. By studying animals in natural habitats rather than in unnatural environments with artificial stimuli his research led him to believe that animals were capable of thinking, that they were conscious and not automata. Publishing his ideas in his 1976 book *The Question of Animal Awareness*, this was – and remains in some quarters – highly controversial and led to accusations of anthropomorphism. However, his seminal work has shaped the discipline of zoology, and scientists have developed the field further by exploring how animals can form concepts and anticipate the actions of others.

Griffin was a prolific writer – as well as a number of books he published over 100 papers between 1938 and 2001. After his retirement in 1986 he lived in Lexington, Massachusetts, where he died in 2003.

Rebecca Thomas

DEVELOPMENT

the 30-second zoology

Development takes an egg and sperm from one generation and turns it into a functioning individual in the next. A single cell becomes trillions in humans – or exactly 959 in the body of *Caenorhabditis elegans*, a nematode worm. Despite these vast differences in numbers, in both worms and people the same processes occur. The first is embryogenesis. Following fertilization the cell divides, increasing the number of cells and allowing differentiation. More than one cell means that they can do different jobs. A cell's relative position in the embryo makes it different from others. Cells on the surface are called ectoderm (outside layer) and the layer on the inside endoderm. Cell movement at gastrulation forms the gut and puts other cells between the outer and inner layers, making mesoderm. From these layers different organs can form, a process called organogenesis. Ectoderm will form skin and the nervous system, while endoderm forms gut tissue. The mesoderm forms muscles and, in vertebrates, the bony skeleton. Development is the result of an interplay between genes specifying a body axis and those identifying specific regions within the embryo and the environment that the cells are in. Specific organizers within embryos secrete proteins to give exact positional identity, meaning that the right body parts form in the right places.

3-SECOND DISSECTION
Development increases cell numbers, differentiates cells into tissues and turns these into organs, but it doesn't stop there; after hatching or birth, further development turns juveniles into adults.

3-MINUTE SYNTHESIS
Development can be seen as an analogy for the process of evolution. The idea that evolution can mould simple single-celled organisms and make complex multicellular ones is often difficult to comprehend: after all, what are the chances? However, as evolutionary biologist J. B. S. Haldane said, 'you did it yourself in 9 months': from a single fertilized egg, you are now sitting reading these words. Evolution and development writ large.

RELATED TOPICS
See also
GENES
page 16

METAMORPHOSIS
page 76

3-SECOND BIOGRAPHIES
J. B. S. HALDANE
1892–1964
British-Indian evolutionary geneticist, biometrician, physiologist and mathematician; introduced the Oparin–Haldane hypothesis on abiogenesis and the primordial-soup theory

HILDE MANGOLD
1898–1924
German embryologist whose exquisite work on the development of amphibians showed how the vertebrate body plan is organized

30-SECOND TEXT
Neil Gostling

This beautiful process takes a single fertilized egg cell and turns it into a multicellular organism with the ability to repeat the whole process again.

A   B   C   D

E   F

METAMORPHOSIS

the 30-second zoology

Metamorphosis is the process

of change that for some groups of organisms results in larvae transforming into very different adult forms. This has the ecological benefit of reducing competition between adults and their offspring for food and habitat resources while also allowing for differences in dispersal. The most recognized forms of metamorphosis are seen in amphibians and insects, but they take radically different routes to adulthood. For example, in frogs it is really only an elaboration of the existing body plan: an outgrowth of legs and the development of lungs, with the resorption of gills and the tail, for the transition from water to land. In flies and other insects, the process is much more extreme. While the fly's head-to-tail axis remains, just about every part of the larval body is broken down, and the adult structures develop from non-functional regions of the larval body. These areas, called imaginal discs, develop at metamorphosis into legs, wings and all of the appendages and anatomical structures necessary for the adult to transition from a life in the soil, exploiting the niche for food and growth, to a world on the wing, looking for a mate. It's at metamorphosis where individuals become sexually mature and can reproduce.

3-SECOND DISSECTION
Metamorphosis occurs when significant physical changes in animal body form occur between life-history stages, such as between the larval and adult stages of butterflies.

3-MINUTE SYNTHESIS
In almost all organisms that partition life history into a larval and adult stage, the ability to reproduce is held back until after metamorphosis. However, one amphibian has found a halfway house between juvenile and adult lifestyles. The axolotl, which inhabits a few miles of river near Mexico City, is a salamander that retains the gills of the juvenile, so cannot survive out of water, but is sexually mature in its juvenile body.

RELATED TOPICS
See also
GENES
page 16

DEVELOPMENT
page 74

3-SECOND BIOGRAPHY
MARIA SIBYLLA MERIAN
1647–1717
German naturalist whose detailed observations of the life cycles of insects were some of the first to describe metamorphosis accurately

30-SECOND TEXT
Neil Gostling

Spiders, invertebrates, butterflies and other insects undergo metamorphosis, as do some tetrapods, including amphibians.

BREATHING

the 30-second zoology

RELATED TOPIC
See also
DEVELOPMENT
page 74

3-SECOND DISSECTION
Breathing is key to aerobic respiration; in the presence of oxygen glucose breaks down and produces carbon dioxide, water and, critically, the energy that powers life.

3-MINUTE SYNTHESIS
While there are four main ways of getting oxygen to where it's needed, some species do things a little differently. Fitzroy River turtles in Australia have lungs but gain around 70 per cent of their oxygen requirement through cloacal respiration (colloquially known as 'breathing through their bums', although, strictly speaking, the cloaca is not the same as the anus), where specialized cells act like alveoli. This allows the turtle to remain under water for several days.

The ability to take air into the body and then exchange oxygen for carbon dioxide is key to animal life itself. Every cell needs oxygen for respiration, and breathing helps get it where it's needed. There are four primary ways of getting oxygen into the body. The simplest sees oxygen diffusing across the surface of the body to the cells beneath, but this is suitable only for small simple animals in moist environments, and larger species require specialized organs. Gills allow animals in aquatic environments to exchange gases. Gills are membranes, often only one cell thick, that allow a rapid exchange of gases, bringing oxygen into contact with the circulatory system. For terrestrial species things are more complicated. Larger invertebrates, such as insects, have tracheal systems. These are networks of air tubes which, as there is no circulatory system, ramify through the insect's body bringing oxygen to the cells where needed. This contrasts with the complexity of lungs found in terrestrial vertebrates, such as humans. Air is drawn through a series of progressively smaller and more branched tubes, starting with the trachea, dividing into bronchi which, in turn, divide into bronchioles and then into tiny sacs, the alveoli, where gas exchange occurs and the circulatory system can carry oxygen to individual cells.

3-SECOND BIOGRAPHY
WILLIAM HARVEY
1578–1657
English physiologist who was the first to provide anatomical evidence for the crucial links between the respiratory and circulatory systems

30-SECOND TEXT
Mark Fellowes

There are several ways of getting oxygen to where it's needed, but what works best is constrained by physiological limitations. Frog eggs rely on diffusion, tadpoles on gills and adults on lungs.

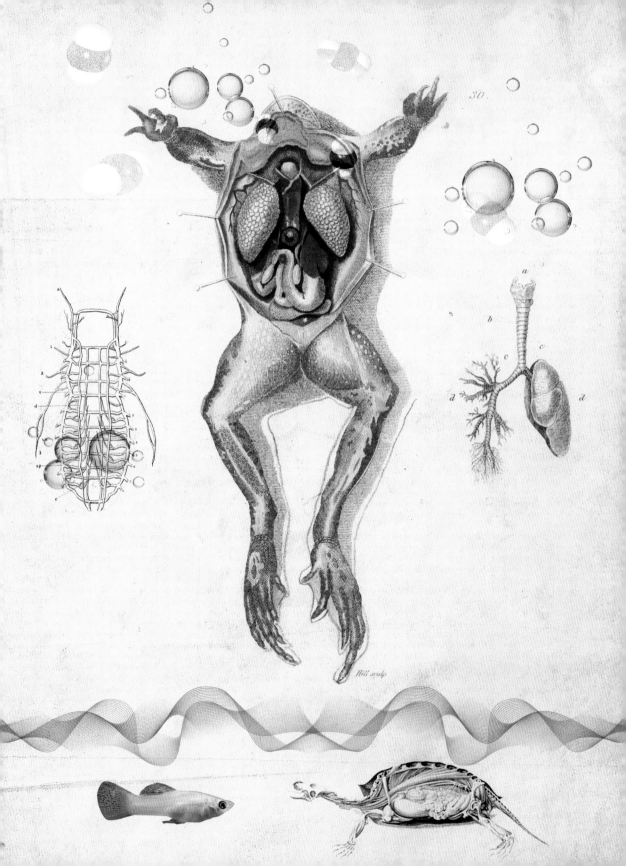

30.

Hill sculp

VISION

the 30-second zoology

The ability to sense light

gives animals access to a high-resolution instantaneous information source with which to assess the world around them. All eyes have opsins, light-sensitive molecules that absorb light of particular energy levels, known by their wavelength. These opsins are packaged into specialized cells (cones and rods) in the retina. We perceive these different wavelengths as different colours, although electromagnetic radiation (light) has no colour in its own right. Instead, colour is created by the brain when it compares the activity of different light-sensitive cells: humans have three types of colour-sensitive cell (cones) while other life forms can have many more, opening up further ways of seeing their environments. The colours that animals can detect depends on which combination of wavelengths they are sensitive to, and many species can see further into the ultraviolet and/or infrared than humans. But animals are not just limited to seeing colour: light can be described both by how much energy it has and by the direction in which the wave propagates, its polarization. This is what polarized sunglasses allow humans to see, but many animals (including crabs and cuttlefish) have this feature built in, giving them the ability to distinguish things that match the background in colour.

RELATED TOPICS
See also
GENES
page 16

PHYLOGENETICS & DIVERSITY
page 22

3-SECOND DISSECTION
Eyes allow animals to see, but not all species see in the same way: humans have three colour channels, dogs have two, birds have four and mantis shrimp have a dozen!

3-MINUTE SYNTHESIS
Eyes have evolved independently many different times to varying levels of complexity, from simple patches of light-sensitive cells found in flatworms to the complex compound eyes of insects and the camera-like eyes of vertebrates and squid. Some animals cannot see colour, and dolphins, whales and seals may see only black, white and shades of grey; others, however, see colours beyond our comprehension. Many birds are even more bright and beautiful than we can ever appreciate.

3-SECOND BIOGRAPHY
HASAN IBN AL-HAYTHAM
c. 965–c. 1040
Scientist born in Basra (present-day Iraq) who investigated how light is received by the eyes and interpreted by the brain

30-SECOND TEXT
James Barnett

Light is all around us, but we humans can only see a small amount of the spectrum, and to other species the world can look quite different.

VENOMS & POISONS

the 30-second zoology

3-SECOND DISSECTION
Both poisons and venoms are toxic chemicals, but poisons must be ingested (you bite it), while venoms are injected (it bites you).

3-MINUTE SYNTHESIS
Not all poisonous species generate their own toxins; some sequester toxins from the plants and animals they eat. Many of the brightest-coloured butterflies and moths gain their chemical defences from the plants they consume as caterpillars. For example, monarch butterflies gain their toxins from milkweed plants and cinnabar moths from ragwort. The colours that make butterflies beautiful to us represent a warning signal to potential predators that says, 'Eat me, and you'll regret it.'

The age-old confusion over terminology: is that snake poisonous or venomous? Both exist, and both produce toxic chemicals that can cause illness or even death, but the distinction lies in whether the trouble results from you biting it or it biting you. Venomous species inject their toxins directly through modified teeth, stings or spines that can be used both for capturing prey and in self-defence; poisonous species excrete or store toxins in or on their bodies, largely to defend themselves from predators or parasites. A huge variety of different chemical toxins is used as both venoms and poisons by different species from jellyfish to frogs to trees. Venoms fall broadly into two categories: those that disrupt the signals between nerve cells (neurotoxins) and those that break down cells (cytotoxins and haemotoxins). Neurotoxins stop nerve signals from reaching muscles, leading to paralysis of vital bodily functions such as breathing and the heart beating. Cytotoxins and haemotoxins break down bodily tissues, disrupting cells and preventing blood from clotting. However, as prey evolve toxicity, predators evolve resistance in an arms race that can result in enormous concentrations of toxins in some species. For example, the rough-skinned newt found in California is so toxic that it could kill several adult humans but is routinely eaten by garter snakes immune to its powerful poison.

RELATED TOPICS
See also
JELLYFISH, CORALS, ANEMONES & MORE
page 36

AMPHIBIANS
page 60

REPTILES
page 62

PREDATION
page 122

3-SECOND BIOGRAPHY
VITAL BRAZIL
1865–1950
Brazilian scientist who developed some of the first snake anti-venoms; his work massively reduced the death rate from snakebites

30-SECOND TEXT
James Barnett

Toxins manufactured by insects, reptiles, amphibians, fish, arthropods and even birds can act as deterrent, defence or deadly weapon.

FLIGHT

the 30-second zoology

Many organisms use air currents to disperse, and some use specialized structures to glide, but true powered flight is special because it allows animals to choose where to go. Powered flight first evolved in insects about 400 million years ago, and much like the metamorphosis from earthbound larvae to flying adults seen today, the transition to the air allowed insects to exploit new niches – and it might also explain the incredible diversity of the group. Flight has also evolved independently from flightless ancestors in three other vertebrate groups – the (now extinct) pterosaurs, birds and bats – each of which has modified its forelimb to make a wing. In pterosaurs the wing was a thin skin membrane supported by the arm and an extremely elongated little finger – indeed, *Pterodactylus* means wing finger. Bird wings incorporate the humerus, radius and ulna and hand, together forming an aerofoil with feathers. Bat wings are formed from the whole hand (the order to which bats belong is Chiroptera, meaning hand wing) the elongated fingers of which have a skin membrane in between. It is hard to say how flight evolved in pterosaurs or bats, because fossil records contain no intermediate forms. Birds, however, have a rich fossil record that shows the transition from small feathered dinosaurs to modern birds.

3-SECOND DISSECTION
Flight, the means to self-propel through the skies, proved advantageous to those organisms that evolved this ability because it opened up new niches and opportunities.

3-MINUTE SYNTHESIS
Charles Darwin predicted that fossils linking the major groups of living animals would be found. In the 1860s a fossil feather was discovered in a quarry in Bavaria, southern Germany, followed by a body fossil covered in feathers. The specimen also had a long bony tail, a jaw with teeth and clawed fingers. It was described as a bird (it had feathers), but it clearly had reptilian, dinosaur features. *Archaeopteryx* was Darwin's predicted link and is described as the first bird.

RELATED TOPICS
See also
NATURAL SELECTION
page 18

PHYLOGENETICS & DIVERSITY
page 22

BIRDS
page 64

METAMORPHOSIS
page 76

3-SECOND BIOGRAPHY
THOMAS HENRY HUXLEY
1825–95
Known as 'Darwin's Bulldog' for his passionate defence of evolutionary theory, Huxley was one of the greatest anatomists of his day and one of the first to suggest a link between birds and dinosaurs

30-SECOND TEXT
Neil Gostling

Flying animals comprise the largest living group (insects) and the largest group of tetrapod vertebrates (birds) on Earth. The bat is the only mammal capable of powered flight.

BEHAVIOUR

binomial nomenclature The classification system whereby organisms are named using a two-word identifier, the first being genus and the second the species.

bombykol The first pheromone to be identified chemically is used by female silk moths to attract males.

eusociality Complex social systems where multiple generations of adults overlap within a single society, and where the majority of individuals do not reproduce to support the reproduction of a queen, as seen in many ants, bees and wasps.

hermaphrodite An organism that possesses both male and female reproductive organs at some stage in its life span. Many invertebrate groups, such as snails, slugs and worms, do not have separate sexes and a number of fish families and some reptile species are also hermaphroditic.

kin selection Where workers are related to the offspring and hence increase their evolutionary fitness by helping to rear siblings.

monogamy Mating system where each male and each female mates with only one partner.

monophyletic group Organisms that included the last common ancestor and all of its descendants, living and extinct.

operant conditioning A learning process by which the expression of a behaviour (the action) is changed by reinforcement or punishment. These can be positive or negative. Positive reinforcement, for example, may involve the use of reward, while negative reinforcement may be associated with avoiding a negative stimulus, such as a loud noise.

pheromone A chemical released into the environment by an animal that acts on the physiology or behaviour of another individual.

phylum (pl. phyla) The first taxonomic division of animals below the kingdom (see taxonomy).

polyandry Mating system in which females mate with multiple males.

polygamy Mating system in which males mate with multiple females.

polygynandry Mating system where both males and females mate with multiple partners.

reciprocal altruism A behavioural trait where an individual acts selflessly to help another at one time but will benefit when the roles are reversed in the future.

sexual dimorphism Where males and females of the same species look or behave differently.

spermatozoa The male sex cells (that is, sperm), the plural of spermatozoon, the mobile single cell that carries the father's DNA to fertilize the egg.

taxonomy The traditional classification system for naming and organizing the animal (and plant) kingdom into groups that share similar qualities. The seven ranks are: kingdom, phylum, class, order, family, genera and species. By contrast, phylogenetic or cladistic nomenclature ties names to a clade, the name for a collection of organisms that group together through the possession of shared characteristics inherited from a common ancestor.

10 June 1929
Born in Birmingham, Alabama, USA

1949/1950
Graduates with a BS then MS in biology from the University of Alabama

1955
Awarded a PhD in biology from Harvard University, Massachusetts, USA

1956–76
Teaches at Harvard, becoming a full professor in 1964

1973–97
Becomes curator in entomology at the Museum of Comparative Zoology, Harvard

1976
Awarded the US National Medal of Science

1979
Wins the Pulitzer Prize for general non-fiction for *On Human Nature*, which deals with the role of biology in the evolution of human culture

1990
Awarded the Crafoord Prize by the Royal Swedish Academy of Sciences

1991
Wins a second Pulitzer for general non-fiction for *The Ants*, co-authored with Bert Hölldobler

1997–
Becomes honorary curator in entomology at the Museum of Comparative Zoology, Harvard

E. O. WILSON

Edward Osborne Wilson was born in 1929 in Birmingham, Alabama, USA. He was precociously interested in science, but a fishing accident when he was seven caused him to lose the vision in one eye, resulting in him becoming focused on 'little things', as he described them, in particular butterflies and ants. By the time he was 18 he was already an expert entomologist, reporting the first colony of invasive fire ants in the USA while still in high school. Wilson attended the University of Alabama before completing his PhD at Harvard in 1955. He remained at Harvard, researching ant taxonomy and evolution and later becoming curator of entomology at Harvard's Museum of Comparative Zoology.

Wilson is perhaps the world's most famous ant expert, particularly in regard to his research on how ants use pheromones to communicate. It was the self-sacrificing nature of ants that helped influence his contribution to a new scientific discipline, sociobiology, which studies animal behaviour from a social perspective. Controversially, Wilson has also applied these perspectives to human behaviour, arguing that it is as much genetically as culturally determined, fuelling debates about the relative levels of nature and nurture in human behaviour.

His most notable contribution to ecology was in his involvement in developing the field of island biogeography, which shows that an island's biodiversity is related to two factors, size and isolation, and how this leads to small isolated islands having fewer species than larger islands near other islands. This theory was tested by dynamiting islands in the Florida Keys to change their size and then removing all insect life with insecticides. By recording the changes in biodiversity over time Wilson was able to show that his theories were essentially correct, and this insight has become the keystone of modern conservation biology.

In the face of the enormous biodiversity losses across the globe, in his later career he has emphasized the importance of conservation, introducing the term 'biophilia' to describe what he argues is humankind's innate, genetically determined need to connect with nature. This has been significant in shaping how conservation biology in the Anthropocene is approached.

Wilson's wider influence is considerable. He has brought to the public's attention fundamental questions about what makes us human, why societies work the way they do and how we must strive to save what remains of the natural world for future generations. He has won numerous prizes, including the Pulitzer Prize (twice) and the Royal Swedish Academy's Crafoord Prize (which recognizes sciences not covered by the Nobel categories) and was named as one of the 25 most influential Americans by *Time* magazine in 1995.

Mark Fellowes

LEARNING

the 30-second zoology

Animals are able to learn from experience, altering behaviours in response to stimuli. At a simple level, this happens through habituation (a reduction in response) or sensitization (an increase in response) to a stimulus. This is non-associative learning, where the presence of a single stimulus is directly linked to a change in behaviour – one example being urban pigeons learning that humans present little threat. More complex is associative learning, where two or more stimuli become associated by the learner. This can be in the form of classical conditioning, where the animal learns to associate a stimulus with a pre-existing behaviour – as seen with Ivan Pavlov's dogs salivating at the ringing of a bell because they had learned to associate the sound with the presence of food – or it can be in the form of operant conditioning, where reward or punishment changes behaviour. This is typical of dog training, where rewards reinforce positive behaviours in response to commands. But other forms of learning are much more complicated. Some species can learn from observing others and copying novel behaviours. Others learn through insight, where experience informs reasoning, a form of learning only found in more intelligent species such as primates and crows. This leads to intricate, adaptable behaviours, as epitomized by our own species.

RELATED TOPIC
See also
NATURAL SELECTION
page 18

3-SECOND DISSECTION
Learning is an ability to modify behaviour and then repeat that behaviour in the future in a manner that may benefit an organism's survival.

3-MINUTE SYNTHESIS
Learning can lead to cultural differences. Chimpanzees use branches, twigs and leaves as tools to help find food. In Uganda, chimpanzees in Kibale use sticks to get honey, but those in Budongo soak up honey with chewed leaves. These cultural differences arose because of chance differences in insight learning, followed by reinforcement in the population from observational learning as skills are passed from one individual to another. Some animals are not so different from humans after all.

3-SECOND BIOGRAPHIES
IVAN PAVLOV
1849–1936
Russian/Soviet physiologist best known for his research on classical conditioning, particularly his work with dogs; he was awarded the Nobel Prize for Physiology or Medicine in 1904

JANE GOODALL
1934–
British primatologist who has shown how extensive tool use is among chimpanzee populations

30-SECOND TEXT
Neil Gostling

Bears learn that where there are bees there is honey but how could bees have learned to construct honeycomb based on a hexagon (the most efficient shape in terms of strength and materials), which appears instinctual?

COMMUNICATION

the 30-second zoology

3-SECOND DISSECTION
Animals communicate
through sights, sounds,
smells, touch and
behaviour to convey
intent, assess mating
potential, warn each
other of danger and to
coordinate group activity.

3-MINUTE SYNTHESIS
We humans communicate
all the time by using
sounds (speech), visual
cues (body language) and
even chemicals (smell).
While communication in
the animal kingdom can
take many different – and,
to us, sometimes surprising
– forms, only a few species
have developed specific
sounds that have been
suggested might function
in a similar manner to
words: bottlenose dolphins
produce unique signature
whistles that identify
individuals, and several
monkeys and birds use
specific alarm calls for
specific predators.

Animals communicate. Their
signals convey information that can influence
the receiver, thus allowing the complex,
complicated richness of animal behaviour to
emerge. Signals may be exchanged through any
of an animal's senses. While humans largely
communicate through sound and gesture, other
species signal each other in ways difficult for
us to imagine, such as the electrical pulses of
knifefish or the chemical cues that make up
pheromones. Pheromones were first isolated
60 years ago when the chemical bombykol –
which female silk moths use to attract males
from great distances – was first described.
While widespread, particularly in insects,
pheromone use by species such as humans
remains highly controversial. Communication
between members of the same species allows
animals to convey and assess one another's
intentions and qualities as a mate or rival as well
as allowing groups to coordinate their collective
behaviour. This may range from relatively simple
olfactory markers left at the edge of a territory
to elaborate multisensory displays and the
complex coordination of group foraging and
defence. Communication between species is
less complex and is usually designed to convey
threats, but it can result in shared benefits,
such as when honeyguides (an African bird) lead
honey badgers (and also humans) to beehives.

RELATED TOPIC
See also
GROUP LIVING
page 100

3-SECOND BIOGRAPHIES
ADOLF BUTENANDT
1903–95
German Nobel Prize-winning
biochemist who identified the
pheromone bombykol in 1959

ALEX THE AFRICAN GREY
1976–2007
A parrot, raised by animal
psychologist Dr Irene
Pepperberg, who developed
a vocabulary of more than
100 words for different objects
and actions

30-SECOND TEXT
James Barnett

*Communication of one
form or another is the
basis of all interactions
between individuals,
groups or species.*

MANAGING CONFLICT

the 30-second zoology

RELATED TOPICS
See also
COMMUNICATION
page 94

ALTRUISM
page 98

GROUP LIVING
page 100

3-SECOND DISSECTION
Conflicts over food, mates and territories arise all the time, and, rather than risk fighting, animals shout, posture and bluff their way through while avoiding physical fights.

3-MINUTE SYNTHESIS
Chameleons are known primarily for their ability to change colour so enabling them to blend into the background and avoid detection by predators, but these diminutive reptiles can also be fiercely territorial, defending their patch from all intruders. Fighting can be dangerous, so, when confronted, chameleons may hiss loudly, inflate themselves, extend exaggerated frills and change colour dramatically from passive green to fiery red. Only if all this fails to scare off a rival will they physically fight.

Animals need access to a few essential things to be successful: territory, food, water and mates. But everyone else out there is after the same, and when there isn't enough to share things can become dangerous. When conflict arises there are two options: fight or flee. Fighting creates the risk of being injured and can waste time and energy, but fleeing means giving up precious resources that are unlikely to be found undefended elsewhere. So perhaps conflict should be managed in a more nuanced fashion. Many species use behavioural displays to pre-empt the outcome of any fight and allow the loser to back down prior to any physical conflict. Mouth gaping, side-by-side struts and displays of horns, bellows or weaponry all allow rivals to size up their opponents and forecast the likely outcome of a fight. Conflict is not always as dramatic as a physical confrontation, but its costs will still need to be mitigated. Competition for resources between individuals of the same species can lead to partitioning of resources and activity patterns so that each exploits their own niche. The same principle extends between species – for example, competitive herbivores specializing towards particular resources or eating leaves on the same plant as other species but at different elevations.

3-SECOND BIOGRAPHY
DIAN FOSSEY
1932–85
American primatologist who transformed the understanding of gorilla behaviour and highlighted how display and communication were used by gorillas to avoid conflict

30-SECOND TEXT
James Barnett

Animals are always in competition with one another, for food, space or mates, but competition does not always need to result in physical conflict.

ALTRUISM

the 30-second zoology

The natural world is full of examples of cooperative animals working together or seemingly acting selflessly for the benefit of others: meerkats take turns giving up food to keep watch while others feed; vampire bats have been known to share food with those who haven't found enough; and ants will sacrifice themselves for the good of the colony. But evolution favours genes that increase survival and reproductive success. Giving precious resources to others would, therefore, appear to go against natural selection, with any *selfless* (altruistic) individuals at risk of being taken advantage of and outcompeted by more *selfish* individuals who keep resources to themselves. This apparent paradox is solved by understanding the indirect benefits of altruistic behaviour: kin selection and reciprocal altruism. Natural selection works not on the species nor even the individual but on the genes within them, and those genes are shared by close relatives. By sharing resources with related individuals and increasing the chances that they can survive and reproduce, acting altruistically can benefit an individual indirectly by increasing the overall representation of their genes in the next generation. But you don't have to be related for altruism to persist. In some social species, being able to remember who's who allows mutually beneficial social networks to persist, with favours granted and returned.

RELATED TOPIC
See also
GROUP LIVING
page 100

3-SECOND BIOGRAPHY
ROBERT TRIVERS
1943–
American evolutionary biologist and sociobiologist who developed the mathematical theory of reciprocal altruism in 1971

30-SECOND TEXT
James Barnett

3-SECOND DISSECTION
Altruism is active and selfless behaviour by one or more individuals to benefit another of their kind or an entire community.

3-MINUTE SYNTHESIS
Reciprocal altruism can be summed up as the 'you scratch my back, and I'll scratch yours' approach. By acting altruistically to those in need, an individual with resources to spare can increase the likelihood of repayment when and if the roles are reversed. Of course, there may be cheats who will attempt to take advantage of such acts of selflessness, so it always pays to keep track of which members of the community are trustworthy and which aren't.

Natural selection favours genes that maximize an individual's survival and reproduction, but being selfish is not always the best way to get the most benefits.

GROUP LIVING

the 30-second zoology

3-SECOND DISSECTION
Living together can allow
groups of animals to
defend themselves better,
raise their young and work
together to find more food
than they could alone.

3-MINUTE SYNTHESIS
Group living has been
taken to extremes by
eusocial species, such
as ants, bees, termites,
some shrimps and naked
mole-rats (the only
mammals that truly live
in this way), showing
the highest levels of
cooperation and sociality
to be found anywhere.
Eusocial colonies,
sometimes numbering
millions of individuals,
work together to raise
offspring, collect food
and build large communal
structures with a clear
division of labour among
groups, or castes, within
the colony.

What do lions, termites, dolphins
and emperor penguins have in common? They all
spend at least part of their lives in groups. But
why would such different animals all evolve to
live socially? Living together makes animals
more conspicuous to predators, allows disease
to spread more easily and puts every individual
in direct competition for resources. Social living
can, however, offer significant benefits. Where
food is scarce, it can allow more efficient
foraging. Groups of predators may be better
able to tackle a wider range of prey, increasing
the reward for each individual and allowing
access to large or well-defended prey that
would otherwise be unavailable. Prey species
benefit from simple safety in numbers: larger
groups have more eyes looking for threats, and,
if attacked, a group is better equipped to fight
off predators. Indeed, safety in numbers works
not just by decreasing the chance that any
individual will be unlucky but, by acting as a
large unpredictable mass, the group can confuse
approaching predators, making the task of
isolating a target more challenging. In eusocial
species, huge numbers cooperate on a level not
seen elsewhere in the animal kingdom. In these
colonies all individuals are closely related, so
working together, even if at personal cost,
maximizes the chances of their genes passing
to the next generation.

RELATED TOPICS
See also
E. O. WILSON
page 92

COMMUNICATION
page 94

ALTRUISM
page 98

3-SECOND BIOGRAPHY
BERT HÖLLDOBLER
1936–
Influential German
sociobiologist and
entomologist who, along
with E. O. Wilson, won the
Pulitzer Prize for *The Ants*,
a book on ant societies

30-SECOND TEXT
James Barnett

*Working together
means sharing:
sharing rewards but
also sharing costs.*

MATING

the 30-second zoology

Mating combines eggs and sperm to create the next generation – but there are many ways of getting there. Species may be monogamous, with just one sexual partner, either one at a time or one for life – although such mating systems are frequently fuzzier than they might at first appear, with extra-pair matings found in even strictly socially monogamous species – or polygamous, with multiple mates at once. In some species males mate with multiple females (polygamy); in others females mate with multiple males (polyandry); and in others again both sexes mate with multiple partners (polygynandry). The mating system favoured depends on how reproductive investment differs between the sexes and how much parental care is required, but males and females may not agree on what strategy is best. The female is most often the limiting sex, as the number of eggs that she can produce is much lower than the number of eggs a male could theoretically fertilize. There are exceptions, however, and where males provide particular care or resources they may act as the limiting sex. This is seen in the red-necked phalarope, a wading bird, where females compete for males who will care for their young; free from the burden of raising chicks, female breeding success is limited only by the number of available males.

RELATED TOPICS
See also
SEXUAL SELECTION
page 104

SPERM COMPETITION
page 106

SEX ALLOCATION
page 108

3-SECOND DISSECTION
Mating is the physical process by which sperm and eggs are brought together, allowing the mixing of parental genes in a way that creates unique new individuals.

3-MINUTE SYNTHESIS
Driven by both natural and sexual selection, mating systems can be highly variable. In hermaphroditic species such as snails, which have both male and female sex organs, you may see competition to decide who will produce eggs and who will produce sperm when they mate. Our own system as humans is no more 'normal' than that of any other species; it is simply the outcome of what helped pass on the most genes in our evolutionary past.

3-SECOND BIOGRAPHY
OLD BLUE
1970–83
By 1980 Old Blue was the last surviving fertile female New Zealand black robin; all black robins today can trace themselves back to this one individual

30-SECOND TEXT
James Barnett

Mating brings sperm and eggs together, but the best system for creating and raising offspring varies greatly.

SEXUAL SELECTION

the 30-second zoology

All species evolve through

natural selection, and genetic mutations that increase survival and reproductive success are more likely to passed on to the next generation. Sexual selection is the subset of natural selection that favours the traits that improve an individual's chance of mating rather than simply surviving. The exaggerated colours, ornaments and behaviours of many species have their roots in impressing mates and defeating rivals. For example, the extravagant feathers of the male peacock's tail are long, cumbersome, expensive to produce and maintain and likely to catch the eye of predators looking for an easy meal, but displaying such a fine train is what shows a peahen that this may be a worthy mate. She wants the best genes for her chicks, genes that will make them successful at finding food and attracting mates of their own. A male who can afford to divert resources away from escaping predators and finding food yet still manage to be successful is likely to pass those high-quality traits on to his offspring, leading to handsome sons and successful daughters. Costly traits are honest signals to prospective mates, highlighting health and success, as weaker mates who may want to cheat will be punished by natural selection.

RELATED TOPICS
See also
NATURAL SELECTION
page 18

MATING
page 102

3-SECOND BIOGRAPHY
AMOTZ ZAHAVI
1928–2017
Israeli evolutionary biologist who pioneered the handicap principle in sexual selection, the idea that sexually selected features impart a cost and act as honest signals of mate quality

30-SECOND TEXT
James Barnett

3-SECOND DISSECTION
Choosing the right mate is important for the future success of any offspring, so many animals go to great lengths to show off their qualities.

3-MINUTE SYNTHESIS
Sexual selection often results in sexual dimorphism (males and females looking and behaving differently), simply because the best strategy for passing on your genes can vary between males and females. Elephant seals show some of the most extreme sexual dimorphism: males can weigh up to 5,000 kilograms (11,000 pounds), while females only reach about 1,000 kilograms (2,200 pounds), and all that extra male bulk is the result of male–male competition for access to females.

Some of the most beautiful and striking animal ornamentation and behaviour has evolved to impress mates and warn rivals.

SPERM COMPETITION

the 30-second zoology

Males and females of many species compete to get access to the most desirable partner. This competition does not necessarily stop once mating is complete, and males may continue to compete through their sperm. When racing to the egg, the healthiest sperm is likely to be one that has the best chance of fertilization. By mating with multiple males, a female can stack the deck towards her eggs being fathered by the male with the best genes. But males are not passive competitors, and they have developed tactics to ensure that they will be the successful father. To increase their chances, males from species with high levels of sperm competition normally have larger testes. Some species have enormous sperm: those of the fruit fly *Drosophila bifurca* can be 56 millimetres ($2^1/_5$ inches) long when unwound, 20 times the length of the flies themselves. Other species may use mechanical or chemical tactics to remove or disable the sperm of rivals, and some insects even plug the female's reproductive tract to stop her from mating again. Females also play a role in determining which sperm are successful. Spermicidal agents in her reproductive tract may kill a large proportion of the spermatozoa, weeding out the weak or undesirable.

3-SECOND DISSECTION
Competition between males continues after mating, as sperm race to make sure they are first to reach the egg; the winner isn't decided until the egg is fertilized.

3-MINUTE SYNTHESIS
Communities of chimps have multiple males competing to mate with multiple females, but gorillas live in groups dominated by a single very large silverback male who guards the females from rivals. Chimps have enormous testes, whereas those of the silverback are relatively tiny. Larger testes mean more sperm, and this is a tactic to overwhelm rivals with sheer numbers – something gorillas do not have to worry about. Intriguingly, human testes are proportionally somewhere in the middle.

RELATED TOPICS
See also
NATURAL SELECTION
page 18

MATING
page 102

SEXUAL SELECTION
page 104

SEX ALLOCATION
page 108

3-SECOND BIOGRAPHY
ANGUS JOHN BATEMAN
1919–96
British geneticist who pioneered Bateman's principle that male reproductive success in most species varies more than that of females

30-SECOND TEXT
James Barnett

Although thousands of sperm may start the race to the egg, only one can be successful.

SEX ALLOCATION

the 30-second zoology

Animals produce offspring for a single purpose: to propagate their genes into the next generation. As humans, we may assume that the (roughly) 1:1 ratio seen between our sexes is simply a property of the equal chance of a sperm bearing an X or a Y chromosome fertilizing an egg, but there is also an evolutionary explanation. If one sex becomes more common, then it will have a reduced relative chance of breeding, so evolution would select for the rarer sex, bringing things back to equilibrium. This assumes that the likely reproductive success of each sex (and the cost of rearing each sex) is equal. It is not always so. In harem-breeding species (such as red deer, where one male guards several females) male reproductive success is highly skewed towards a small number of dominant males, whereas female reproductive success is not limited by mate availability. A son represents a high-stakes bet with only the highest-quality sons likely to breed, but, when they do, they sire many offspring. Daughters are almost guaranteed to reproduce but are limited in the number of offspring they can produce. Not surprisingly, smaller, younger mothers tend to produce more female offspring, thereby maximizing their chances of spreading their genes to the succeeding generation.

RELATED TOPICS
See also
NATURAL SELECTION
page 18

SEXUAL SELECTION
page 104

SPERM COMPETITION
page 106

3-SECOND BIOGRAPHIES
RONALD FISHER
1890–1962
British statistician who developed a theory of why sex ratios in most animals are close to 1:1

WILLIAM D. HAMILTON
1936–2000
British evolutionary biologist, much influenced by Fisher, who revolutionized our understanding of extraordinary sex ratios

30-SECOND TEXT
James Barnett

Animals reproduce to increase the proportion of their genes in the next generation but, depending on circumstance, sons and daughters may differ in their chances of reproducing.

3-SECOND DISSECTION
When animals produce offspring, males and females may have different benefits or costs, and many species can alter the sex ratio of their progeny.

3-MINUTE SYNTHESIS
Sex ratios may at first appear to be random, but in certain species females have the ability to skew the sex ratio of their offspring. In African wild-dog groups, for example, the dominant female is assisted by her sons who stay in the pack and help to defend and feed the next generation. In small groups where more helpers would be more useful, dominant females produce more male-skewed sex ratios in their pups.

ECOLOGY

abiosis The condition of being of non-living origin.

Allee effect For many species being at low density is beneficial, reducing competition and allowing faster reproduction. However, on occasion, species benefit from neighbours, to overcome host defences or to provide many eyes to look out for enemies. Here at low densities they do less well than would be expected. Some suggest that the very rapid decline of the passenger pigeon was in part caused by the Allee effect.

allogenic engineers Ecosystem engineers that alter their environments through their behaviours, such as digging through soil or by building dams.

aposematic coloration Bright coloration that warns predators that the potential victim may be poisonous, dangerous or distasteful.

autogenic engineers Ecosystem engineers that alter their environments by their presence – trees or corals are examples.

bacterial symbionts Bacteria that rely upon their host for survival, and from which the host benefits. Typically these bacteria may help animals digest their food, providing essential nutrients which

their host cannot extract from their food or synthesize themselves.

coevolution The reciprocal evolutionary interaction between two or more species, with a change in one selecting for adaptation in the other, which again leads to a continual cycle of change and response.

commensal species Species that take resources from another without causing it loss.

competitive exclusion principle The concept that no two species living in the same space can have identical ecological requirements; one will always outcompete the other.

ectoparasite A parasite that dwells on the outside of its host, such as fleas and lice.

endoparasite A parasite that lives inside its host, such as tapeworms and viruses.

energetics The flow of energy through an ecosystem, usually starting with sunlight and ultimately dissipated as heat, after passing through a food chain as one individual eats another.

facultative relationships Interactions between species in a mutualism where both will survive in the absence of the other, even if they would do better together. This is usually contrasted with obligate relationships, where both species are required for each other's survival, such as some specialist plant–pollinator interactions.

inter/intraspecific competition Competition is the struggle between individuals for resources such as food or hiding places. This can be with members of the same species (intraspecific) or with other species (interspecific).

monotremes A group of mammals that lay eggs, but also feed their young milk. The only extant monotremes are the platypus and four echidna species.

parasitoids Typically insects (usually wasps) that lay their eggs on or inside other insects. The larvae act as parasites, but eventually kill their hosts when they emerge as adults.

primary productivity The synthesis of complex organic compounds from carbon dioxide and water, mainly through the process of photosynthesis. In essence primary productivity is the amount of plant biomass produced in a given area.

rewilding The restoration of an area to a condition more as it once was before the influence of people. In recent years, rewilding efforts have focused on reintroducing large mammals such as wolf, beaver and lynx to regions where they had been exterminated.

scramble competition When a limited resource is equally available to all individuals, so that the amount of food available to each declines as the carrying capacity of the resource is approached. By contrast, in contest competition, some individuals dominate the resource and always get enough, while others lose.

speciation The process by which new species are formed when genetically isolated.

trophic cascades Events that occur in food chains where a change in the abundance of a predator species flows through the system, changing the numbers of herbivores, and therefore indirectly affecting the abundance of plants. The concept of trophic cascades explains why the loss of top predators in ecosystems is particularly damaging.

vector species Animals that pass a disease from one host to another, such as mosquitoes.

Wallace's Line An imaginary divide between Australasia and Asia, which demarcates regions with very different flora and fauna.

zoonoses Diseases passed from animals to people, such as Ebola, from bats, and anthrax, from cattle.

8 May 1936
Born in Sydney, Australia

1956/1959
Graduates with a BSc
in chemical engineering
and theoretical physics
followed by a PhD in
theoretical physics from
the University of Sydney,
Australia

1959
Becomes a lecturer in
applied mathematics
at Harvard University,
Massachusetts, USA

1962
Becomes a lecturer at
the University of Sydney
and is later promoted to
professor of theoretical
physics

1973
Appointed professor of
zoology, Princeton
University, New Jersey,
USA

1979
Awarded a fellowship of
the Royal Society, UK

1988–95
After relocating to the UK
serves as a Royal Society
research professor, jointly
at the University of
Oxford and at Imperial
College, London

1992
Elected as a foreign
associate of the United
States National Academy
of Sciences

1995–2000
Serves as chief scientific
adviser to the UK Office
of Science and
Technology

1996
Appointed Knight
Bachelor for services
to science; awarded
the Crafoord Prize by the
Royal Swedish Academy
of Sciences for
contributions to the
study of biodiversity

2000–5
President of the Royal
Society

ROBERT MAY

Robert McCredie 'Bob' May

was born in 1936 in Sydney, Australia. As a boy he excelled in science, later attending the University of Sydney to study chemical engineering and physics before completing a PhD there in theoretical physics. This was followed by postdoctoral work at Harvard University before a return to Sydney where he became professor of physics. It was at this time that May became interested in ecological questions, in part influenced by the movement for social responsibility in science.

While a successful physicist, May had his greatest impact in population ecology. His first ecological work investigated ideas of how the stability of ecosystems is influenced by their complexity. Many had argued that increased numbers of species in an ecosystem resulted in greater resistance to change, but May's mathematical approach showed that this was not necessarily so and that, counterintuitively, simpler systems were more robust. This opened a hugely important debate which has helped explain what it is about patterns of species interactions in nature that allows complex ecosystems to persist.

May took a sabbatical in the UK, where he spent time at Imperial College with noted ecologist Sir Richard Southwood, then went to Princeton, where in 1973 he was appointed professor of zoology. While at Princeton May made many contributions to mathematical population ecology, notably highlighting the effects of chaos on population dynamics. This work was hugely influential, helping link simple ecological processes of reproduction and mortality with complex and often unpredictable outcomes. May later turned to questions of disease transmission, estimating how many species are on Earth, conservation and over-harvesting, bringing a fresh mathematical viewpoint to subjects that are of immense importance. Later in his career he applied these ideas to banking, using ecological perspectives to provide institutions such as the Bank of England with insights into how global financial crashes might best be avoided in the future.

In 1988 May moved to Britain, holding a Royal Society professorship jointly at Imperial College and Oxford University. He has won many prestigious awards, including the Royal Swedish Academy's Crafoord Prize, which recognizes disciplines not covered by Nobel categories. May was knighted in 1996 and appointed to the House of Lords in 2001, choosing the title Baron May of Oxford.

May's influence has been enormous. He was chief scientific adviser to the UK government and head of the Office of Science and Technology and president of the Royal Society. Currently he is an Emeritus Professor of the University of Oxford and Fellow of Merton College, Oxford.

Mark Fellowes

BIOGEOGRAPHY

the 30-second zoology

3-SECOND DISSECTION
Species distribution is explained by a complex series of interacting factors, from speciation (how genetically isolated animals evolve into distinct species) and plate tectonics to energy flow and ecology.

3-MINUTE SYNTHESIS
Between the Indonesian islands of Bali and Lombok is 'Wallace's Line', the most famous of biogeographical boundaries. While only 35 kilometres (22 miles) apart, these islands belong to different biogeographical zones. East of the line are the animals of Australia, the marsupials and monotremes, cockatoos and kangaroos; west of the line are the placental mammals, Asia's primates, pheasants and big cats. This boundary demarcates two regions with differing evolutionary histories driven by changes in sea levels and isolation.

There are estimated to be more than 10 million species, but they are not spread equally across the world. Biogeography attempts to explain why this is so. One obvious answer is that humankind has hugely simplified ecosystems, and fewer species are found where people have had the greatest impact. Once we take that into account we still see broad patterns in species richness. Understanding these helps zoologists uncover the ecological and evolutionary processes that cause these differences. One pattern is the latitudinal gradient in numbers of species: the closer to the equator we are, the more species we find. This is largely the result of energy levels; increased levels of solar radiation and rainfall found as we move away from the poles allow an increase in primary productivity (plant growth), which supports more herbivores and, in turn, more predators and parasites, leading to increased diversity. Another simple rule is that of area and isolation. All else being equal, for patches or islands of a given size, those more isolated will have fewer species; for patches of equal isolation, larger patches will have more species. Thus smaller more isolated islands will have fewer species. Isolation limits which species find islands as they disperse, and size limits the number of niches that can be exploited.

RELATED TOPICS
See also
ALFRED RUSSEL WALLACE
page 14

E. O. WILSON
page 92

COMPETITION
page 118

PREDATION
page 122

MUTUALISMS
page 126

KEYSTONE SPECIES
page 128

3-SECOND BIOGRAPHY
GEORGE EVELYN HUTCHINSON
1903–91
British ecologist whose seminal 1959 paper 'Homage to Santa Rosalia', picked out energetics as a key factor explaining why biodiversity declines with distance from the equator

30-SECOND TEXT
Mark Fellowes

Biogeography explains why species are where they are.

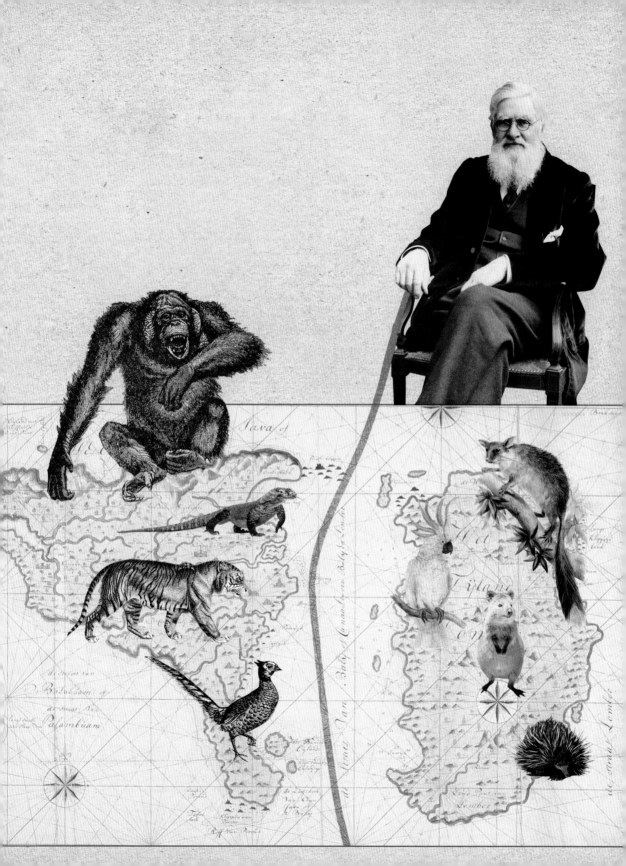

COMPETITION

the 30-second zoology

In the absence of predators

and disease, animal population sizes are limited by the availability of resources. At low sizes populations can grow exponentially, as resources are easy to find – we witness this when pests such as locusts invade a new region. As their numbers reach plague proportions, however, individuals have to compete for food, slowing down how quickly they can grow and reproduce, until the population reaches its carrying capacity, where birth rate is balanced by death through lack of resources. Competition does not have to be fair; it can have winners and losers (contest competition), where a dominant individual gains resources and others lose out. In contrast, with scramble competition everyone loses. Here competitors are more equal and so all take a proportion of the food available, leading to starvation when the resources aren't plentiful enough to maintain every individual. A key tenet of competition theory is that it is density dependent, where increased numbers results in increased competition. It is worth noting that this isn't always the case, however. The Allee effect shows that at low densities populations can do less well than might be expected, as larger numbers working together can help access resources that would otherwise be inaccessible. Sometimes it's better to have a few neighbours than none at all.

3-SECOND DISSECTION
Resources, and competition for those resources, limits population growth; with no competition our planet would be covered by the offspring of a single *E. coli* in less than two days.

3-MINUTE SYNTHESIS
Competition for food, space and mating opportunities can be within (intraspecific) or between (interspecific) species. One hugely influential idea in interspecific competition is that of the competitive exclusion principle, which suggests that when two species are competing for the same limiting resource they cannot coexist; one will eventually win, excluding the other. However, we rarely see this in nature, as who wins is not often determined by any one single factor. The world is a complicated place.

RELATED TOPICS
See also
HERBIVORY
page 120

PREDATION
page 122

MUTUALISMS
page 126

3-SECOND BIOGRAPHIES
WARDER CLYDE ALLEE
1885–1955
American ecologist, after whom the Allee effect is named

GEORGY FRANTSEVICH GAUSE
1910–86
Soviet biologist who developed the competitive exclusion principle

30-SECOND TEXT
Mark Fellowes

Competition is everywhere. Without it, there are few limits to population growth.

HERBIVORY

the 30-second zoology

3-SECOND DISSECTION
Reliant on a plant-only diet,
herbivores are trapped
between predators (some
of which are deliberately
attracted by plants to
help defend them) and
poor-quality food.

3-MINUTE SYNTHESIS
Herbivores have evolved
a wide range of means to
counter the defences
plants put up against them.
Most rely upon bacterial
symbionts (bacteria living
symbiotically within the
herbivore) to aid the
digestion of plant material,
and these make up a
significant proportion of
their microbiome. Others,
such as proboscis monkeys
and ruminants such as
cattle, have more
complicated, slow digestive
systems. Others just take
their time; there's a very
good reason why sloths
are slothful.

Why is the world green? This
superficially simple question was posed in
1960 by three ecologists who wondered why
herbivores have not completely over-exploited
the plants they consume. They argued that
plants survive because predators limit herbivore
numbers, a top-down process. This is certainly
true in some circumstances – for example,
where sea otters were hunted to extinction,
sea urchins flourished and destroyed kelp
forests. Others suggested that predators are
less important than they might appear. Plants
are not just passive victims waiting to be
consumed; they have a range of physical and
chemical defences that mean that they are
just very poor-quality food. Herbivores have
to overcome spines and thorns, saps and latex,
before they can even begin to feed; and when
they do, they have to deal with silica (which
wear down mandibles) and a complex chemistry
which can both poison the herbivore and further
reduce the nutritional quality of the plant by
binding to essential nitrogen compounds and
rendering them inaccessible. So the world may,
in fact, be green because herbivore numbers
are limited by poor nutrition. Whether it's
top-down or bottom-up processes keeping the
world green, herbivory has driven the evolution
of plant chemistry, giving humanity some of
its most utilized stimulants (caffeine), flavours
(chilli) and drugs (cocaine, heroin).

RELATED TOPICS
See also
COMPETITION
page 118

PREDATION
page 122

MUTUALISMS
page 126

KEYSTONE SPECIES
page 128

3-SECOND BIOGRAPHIES
NELSON HAIRSTON SR
1917–2008
FREDERICK SMITH
1920–2012
LAWRENCE SLOBODKIN
1928–2009
American zoologists and
ecologists who popularized the
concept of trophic cascades, in
which predators benefit plants
by removing herbivores

30-SECOND TEXT
Mark Fellowes

*Being a herbivore isn't
easy. Plants may
appear defenceless,
but they hide a complex
chemistry to keep
enemies at bay.*

PREDATION

the 30-second zoology

Predators kill their prey. Both killers and killed interact in ecological and evolutionary time, coevolving defence and counter-defence. Some of the most glorious animals alive have been shaped by their lives as hunter or hunted, with evolution honing traits that increase success. Predators become fast, masters of camouflage and concealment and heavily weaponized with claws, teeth or venom. Each is an adaptation that increases the success of predation, and, in turn, this evolutionary pressure has selected prey that are fast, agile, hidden, armoured, vigilant and sometimes poisonous, with bright aposematic colours that warn predators that they are not good to eat. Some prey cooperate, using signals to alert others to the presence of a threat, while others bluff, mimicking larger or more dangerous animals. This escalation of traits that affect the likelihood of eating or being eaten is essentially an arms race. Some argue that selection pressure is asymmetrical, with stronger selection for escape. This is the 'life–dinner principle'. Prey that fail to escape lose their lives and leave no offspring, whereas a hunter that fails to kill just loses a meal. This means that there may be limits to the pace of coevolution between predators and prey, but it also ignores the fact that missing enough meals also ultimately leads to death.

3-SECOND DISSECTION
For both hunter and hunted, predation is a fact of life – and this includes predation of predators all the way to the top of the food chain.

3-MINUTE SYNTHESIS
Predators and prey do not always coexist. Introduced predators have wreaked havoc on biodiversity in many parts of the world. The domestic cat is widely considered to be one of the most damaging invasive species, having caused 63 known recent extinctions of bird, mammal and reptile species. The best-known example is Lyall's wren, a flightless New Zealand bird species that was lost to the world through cat predation in 1894.

RELATED TOPICS
See also
COMPETITION
page 118

HERBIVORY
page 120

MUTUALISMS
page 126

KEYSTONE SPECIES
page 128

3-SECOND BIOGRAPHY
JOHN RICHARD KREBS
1945–
British ornithologist, who, along with Richard Dawkins (1941–), popularized the concept of the 'life-dinner principle'

30-SECOND TEXT
Mark Fellowes

Predators and prey are entwined in a dance of natural selection and survival.

PARASITISM

the 30-second zoology

3-SECOND DISSECTION
Parasites are species that
live on or in other species,
where the parasite benefits
by consuming nutrients
from the host, and the
host, in turn, is harmed.

3-MINUTE SYNTHESIS
Parasites are hidden,
but their effects are
everywhere. A major threat
to human health comes
from zoonoses, diseases
which jump the species
barrier from animals to
humans. Diseases such as
HIV and Zika are thought
to be caused by viruses
moving from monkeys to
people. The Spanish flu
pandemic of 1918–20 killed
up to 100 million people,
and, more recently,
enormous efforts have
been spent monitoring the
spread of bird flu through
fears that another
pandemic could strike.

Whether ectoparasitic (such as
lice, fleas or fungi) or endoparasitic (such as
flukes, tapeworms, bacteria and viruses) one
thing common to all parasites is that the
relationship with the host only goes one way.
Their life cycles vary greatly in complexity,
but each life stage is exquisitely adapted to
transmission (infecting new hosts, often
involving the use of vector species) and survival
(avoiding the host's immune or behavioural
defences). Parasites are ubiquitous, and almost
everyone will at some point have been a host,
even if only to the innocuous eyelash mite.
Most parasites cause relatively minor damage to
healthy hosts, but the consequences of infection
can be severe. While we are familiar with the
effects of parasites on humans – malaria, for
example – they can also have disastrous effects
on animal populations. One of the greatest
threats to amphibians comes from the chytrid
fungus, which threatens many species across
the world with extinction. The fungus may have
been spread from Africa with exports of the
African clawed frog, which was used between
the 1930s and 1950s for pregnancy testing.
Parasites can, however, be beneficial. Cassava
is threatened by the invasive cassava mealybug,
but the release of specialized insect parasites
(parasitoids) has helped bring it under control,
saving the staple crop of millions of people.

RELATED TOPICS
See also
COMPETITION
page 118

MUTUALISMS
page 126

3-SECOND BIOGRAPHIES
ALPHONSE LAVERAN
1845–1922
French doctor who suggested
that a protist caused malaria;
he was awarded the 1907 Nobel
Prize for Physiology or Medicine

CARLOS FINLAY
1833–1913
Cuban doctor who suggested
that mosquitoes were vectors
of diseases such as malaria

TU YOUYOU
1930–
Chinese chemist who showed
that an extract from a plant
used in Chinese traditional
medicine, artemisinin, was
effective in controlling malaria;
she received the Nobel Prize for
Physiology or Medicine in 2015

30-SECOND TEXT
Mark Fellowes

*Parasites are the most
intimate of species,
supremely adapted to
their host's biology.*

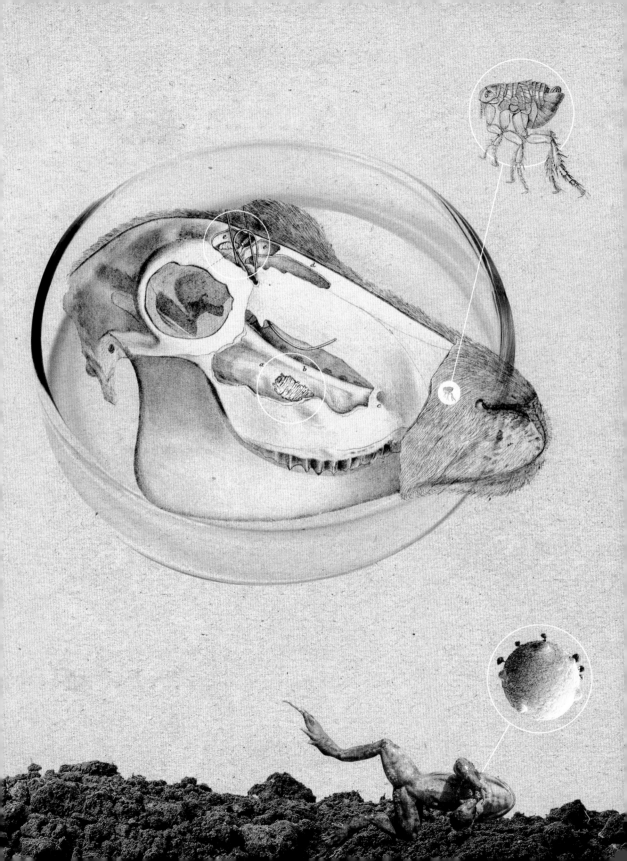

MUTUALISMS

the 30-second zoology

3-SECOND DISSECTION
Not all interactions in
nature are brutal or one
sided; some species have
evolved over time to
cooperate for their
mutual benefit.

3-MINUTE SYNTHESIS
Without mutualistic
interactions our society
would not exist.
Mutualisms drive the
production of crops (with
pollinators) and livestock
(the herbivores we eat and
milk need bacteria to digest
their food plants). Some
scientists argue that,
because of artificial
breeding for traits such as
productivity and docility
over time, most of the
animals and plants we
currently farm would not
survive without us, and we,
in our turn, rely on them
for our own survival.

'Nature red in tooth and claw',
taken from Tennyson's poem *In Memoriam*,
are words that have often been used to describe
the reality of nature: you eat or you are eaten;
you win or you die. This view dominated ecology
and evolutionary biology for many decades
until that perception was whittled away by the
recognition that cooperation between species,
known as mutualism, is widespread in nature.
While this cooperation arises from self-interest,
with both species benefiting, over evolutionary
time some mutualisms have become obligate,
where survival of the interacting species
requires the presence of both. In tropical
forests, fig trees are keystone species, providing
resources for numerous other species. The
existence of each of the approximately
900 fig species is predicated on the presence
of a pollinating fig wasp, with almost every
tree species having a single specialist pollinator
species. Neither fig tree nor fig wasp would
survive without the other. Other mutualisms
are facultative, where one interacting species
can survive in the absence of the another but
will benefit if they are together. In many parts
of the world, species of true bug (such as
aphids) are tended by ants. The ants protect
the insects from predation while benefiting
from the honeydew (a sugar-rich secretion)
produced by the bugs.

RELATED TOPICS
See also
COMPETITION
page 118

HERBIVORY
page 120

PREDATION
page 122

KEYSTONE SPECIES
page 128

ECOSYSTEM ENGINEERS
page 130

3-SECOND BIOGRAPHY
PIERRE-JOSEPH VAN
BENEDEN
1809–94
Belgian zoologist who, in 1876,
introduced the term mutualism

30-SECOND TEXT
Mark Fellowes

*Sometimes Nature
allows both sides to win
the game of life.*

KEYSTONE SPECIES

the 30-second zoology

3-SECOND DISSECTION
Out of all proportion to their abundance, keystone species determine the structure and stability of ecosystems, and their loss can cause lasting damage to biodiversity.

3-MINUTE SYNTHESIS
Keystone species are immensely important within their environments, but throughout history humans have disproportionately removed them from ecosystems. We have a tendency to target large predators, whose presence prevents a small number of strong competitors from dominating ecosystems. If we are to protect or rebuild ecosystems, rewilding has to include the return of keystone species. The enormous and beneficial effect of a handful of wolves reintroduced to Yellowstone National Park stands as a testament to this approach.

A keystone holds an arch together; without it, everything collapses. In ecosystems the same critical role is provided by keystone species, a term defined by Bob Paine in 1966 after studying rocky-shore intertidal communities in Washington State, USA. Paine found that removing a species of starfish resulted in a simplification of the natural community with almost half of species ultimately disappearing, because, by predating mollusc species that would otherwise have outcompeted others, the starfish maintained species diversity. Since his discovery, many examples of keystone species have been found. Perhaps the most impressive example is the grey wolf, following its reintroduction into Yellowstone National Park, Wyoming, USA. The wolf had been exterminated in the region by the 1930s, allowing populations of their prey species, elk, to grow rapidly. Elk overgrazed saplings, and this had a cascade of negative effects, including a reduction in songbird diversity, fewer beavers and simplification of river flow and increased erosion. The wolf's reintroduction in the 1990s helped turn back the clock, as predation directly (elk numbers) and indirectly (elk behaviour) reduced levels of tree browsing. In consequence, songbird diversity and beaver numbers increased, and rivers are returning to their former complexity. The surprising thing? There are only around 100–150 wolves in Yellowstone, an area of almost 9,000 square kilometres (3,500 square miles).

RELATED TOPICS
See also
COMPETITION
page 118

HERBIVORY
page 120

PREDATION
page 122

MUTUALISMS
page 128

ECOSYSTEM ENGINEERS
page 130

3-SECOND BIOGRAPHY
ROBERT 'BOB' T. PAINE
1933–2016
American zoologist who developed the concept of keystone species, influencing generations of ecologists and conservation biologists

30-SECOND TEXT
Mark Fellowes

Top predators help ecosystems to function; their presence benefits biodiversity.

ECOSYSTEM ENGINEERS

the 30-second zoology

Animals are not simply passive residents in an abiotic world; they shape the world, for themselves and other species. All species make small changes to their surroundings, but in 1994 Clive Jones and colleagues suggested that some organisms have inordinately large effects on their environments and consequently for the species around them. They called such species ecosystem engineers. These species can change one element of the abiotic environment into another by mechanical means (allogenic engineers) or by their own growth and presence (autogenic engineers). Beavers are allogenic engineers. By building dams across rivers they slow the passage of water, creating new ponds and lakes. In doing so they provide rich habitats for fish that prefer still waters, reduce erosion and flooding downstream and enhance nutrient capture for the multitudes that live in the sediment. By changing the nature of the ecosystem, beavers benefit many other species. Corals are autogenic engineers. The physical structure of their living and dead forms creates a habitat for massed schools of fish, making coral reefs some of the most diverse places on Earth. Ecosystem engineers are all around us: moles and earthworms turn soil over in our gardens, woodpeckers drill holes in trees, cattle and deer leave tracks which fill with water. Each creates a changed environment for new species to exploit.

3-SECOND DISSECTION
Ecosystem engineers are species that modify their local environments, greatly changing the availability of food and habitats that species other than themselves can exploit.

3-MINUTE SYNTHESIS
The greatest ecosystem engineer of all is humankind. No other single species has altered its environment as we have, from the composition of the atmosphere to plastic pollution at the greatest ocean depths. While nature's ecosystem engineers alter environments in a way that benefits many species, we have engineered the world to benefit a handful of domesticated and commensal species. Most ecosystem engineers increase the diversity of life in their communities, in contrast we simplify and homogenize the living world.

RELATED TOPIC
See also
HABITAT LOSS
page 140

3-SECOND BIOGRAPHIES
CHARLES DARWIN
1809–82
British naturalist who studied the rate at which earthworms turn over soil by measuring how heavy stones sank into the ground over time, an early example of thinking about ecosystem engineers; his book *The Formation of Vegetable Mould, through the Actions of Worms* was his last

MOSHE SHACHAK
1936–

SIR JOHN LAWTON
1943–

CLIVE G. JONES
1951–
Ecologists who wrote the 1994 paper 'Organisms as Ecosystem Engineers', which introduced the idea of ecosystem engineering

30-SECOND TEXT
Mark Fellowes

Life has moulded the planet, providing the oxygen in the air and the soil on the ground.

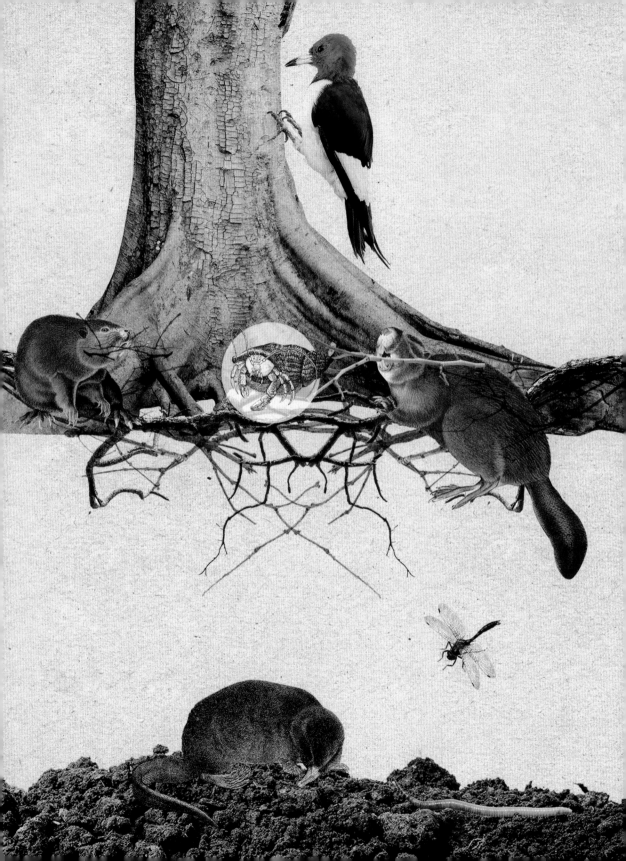

CONSERVATION & EXTINCTION

CONSERVATION & EXTINCTION
GLOSSARY

Anthropocene The proposed term for the current epoch or geological time period of the Earth, beginning with the point from which human influence altered the climate and geological, hydrological, biospherical and other environmental processes.

bioaccumulation The gradual accumulation of contaminants within or on an organism.

biotic homogenization The process by which species found within two or more areas become similar over time because of species invasions and extinctions.

bush meat Term for meat obtained from hunting or trapping non-domesticated animals, a widespread practice that threatens native wildlife diversity.

bycatch The unintended fish or other marine animals that are caught, especially by commercial fishing fleets using longlines, trawlers, gill nets and purse-seine nets is known in the fishing industry as bycatch. It can include different species entirely, the wrong sex, or juveniles of the target species. Marine life can include turtles, dolphins and seabirds that are caught with the intended species, only to be discarded overboard.

disease vector An organism that carries and transmits an infectious pathogen into another organism. Mosquitoes and ticks are vectors for many diseases.

ecosphere The terms for all the ecosystems on our planet.

maximum sustainable yields The largest number of individuals that can be harvested from a population without causing declines over time.

nature deficit disorder Humans are losing their connection with nature. This detachment, or so-called 'nature deficit disorder', is negatively affecting our health and well-being and has implications for conservation worldwide. With more people living in urban areas, our interactions are shaped by the animals that are able to live alongside us. Many people choose to make their gardens and outside spaces more wildlife friendly. Simple acts like putting up a bird feeder can help increase people's daily interactions with wildlife.

phenological asynchrony A mismatch in the timing of lifecycle events between species that rely on each other to survive.

reproductive isolation Mechanisms which prevent members of a species from breeding.

rewilding The restoration of an area to a condition more as it would have been before the influence of humans. In recent years, rewilding efforts have focused on the reintroduction of large mammals such as wolf, beaver and lynx to regions where they had been exterminated.

speciation The process by which new species are formed.

tragedy of the commons Term describing the exploitation and depletion of a natural resource by individuals to their own advantage without consideration the common good.

10 September 1944
Born in Doylestown, Pennsylvania, USA

1966
Graduates with bachelor's degree in biology from Hood College, Maryland, USA

1985
Receives her PhD in anthropology from City University of New York

1986
Travels to Madagascar to search for the greater bamboo lemur, believed at the time to be extinct

1991
Following extensive work, Ranomafana National Park is established in Madagascar

2003
The Centre ValBio, a research institution that aims to promote environmental conservation in Madagascar and throughout the world, is founded

2014
Becomes the first woman to receive the Indianapolis Prize for Conservation

PATRICIA WRIGHT

There are many ways to become interested in science, but Patricia Wright's journey is somewhat unusual. As a mother at home in Brooklyn, New York, in the 1960s, she owned two nocturnal owl monkeys. She was fascinated by their behaviour and wanted to learn more about these primates in the wild; so, with her husband and daughter in tow, she embarked on an expedition to the jungles of Peru to study them. Nearly a decade later she received her doctorate before continuing her journey into conservation science.

Wright later set herself the challenge of trying to locate a species that many considered to be extinct. She travelled to Madagascar, a country filled with endemic species, on the hunt for the greater bamboo lemur. She not only rediscovered the greater bamboo lemur but she also found another new to science, the golden bamboo lemur. As amazing as these discoveries were, Wright was shocked by the devastation she saw around her. Madagascar had already lost so much of its natural habitat, but with logging threatening the survival of these bamboo lemurs as well as many other animals, she knew she must act. Collaborating with local people and the Madagascan government, Wright worked to gain support and raise funds

to set up the Ranomafana National Park. In return for not exploiting the resources in the forests, local people were provided with schools and healthcare facilities as well as new employment opportunities within the park. The park now receives over 100,000 visitors each year, bringing much-needed money into the local economy.

With this incredible area given formal protection, Wright set about establishing Centre ValBio, a research facility positioned next to Ranomafana National Park. The centre has enabled exciting research to be carried out throughout Madagascar, but it is also committed to reducing poverty in the local area to allow for sustainable use of natural resources. With local conservation clubs and the establishment of ecotourism in the region, the local people are gaining an understanding of the value of conservation.

Patricia Wright's conservation message is also being felt at an international level. Each year Centre ValBio welcomes students from across the world in Study Abroad programmes, and Patricia herself and the research being carried out at the centre have been showcased in numerous media outlets, including the 2014 film *Island of Lemurs: Madagascar*.

Rebecca Thomas

CLIMATE CHANGE

the 30-second zoology

Climate change is perhaps the greatest global threat to both biodiversity and humanity. The world's average surface temperature has increased by 0.9°C (1.62°F) since the late nineteenth century – five of the warmest years on record have taken place since 2010 – and by the end of this century conservative estimates predict a further increase of 2°C (3.6°F). Increasing levels of carbon dioxide from human activities are leading to ocean acidification and rising temperatures, causing sea-level rises, changes in rainfall, wildfires, intense storms and glacial retreat. These changes are occurring at such a pace that many species are unlikely to be able to adapt. Vegetation patterns have already changed. In temperate regions trees come into leaf weeks earlier than they once did, and temperature change also affects the insects that feed on them. But the migration behaviour of birds such as pied flycatchers has not changed, meaning that they may arrive late for peak caterpillar numbers. This phenological asynchrony means that they are out of step with their food resource, reducing their chances of successfully rearing young. Not all species will lose out, however. Models predict that many diseases will move into new regions as rising temperatures open up suitable habitats for their vectors.

RELATED TOPIC
See also
HABITAT LOSS
page 140

3-SECOND BIOGRAPHY
WALLACE SMITH BROECKER
1931–2019
American geophysicist, dubbed the 'grandfather of climate science', who popularized the term global warming

30-SECOND TEXT
Rebecca Thomas

3-SECOND DISSECTION
If we do not act fast to reduce carbon emissions climatologists predict a tipping point will be reached from which it will take centuries for the damage to be undone.

3-MINUTE SYNTHESIS
Climate change is taking place so rapidly that many species are not able to adapt to the changing conditions in which they find themselves. With this threat we need to take a landscape view when we consider the conservation of species. The world's natural habitats have become increasingly fragmented, and we need to connect these areas up to allow species to alter their distributions if they are to survive the global impact of climate change.

Humans are driving climate change through global warming and its effects are visible in every part of the world.

HABITAT LOSS

the 30-second zoology

Humans are transforming vast swathes of our planet's surface at a pace unlike any other seen in recent geological time, and habitats are being altered so dramatically that habitat loss is now the greatest cause of global species extinction. Throughout human history we have changed the land, mostly for small-scale agriculture, but more recently we have created vast mega-cities and industrial-scale agriculture so that today over half of the Earth's land surface is no longer natural. This has been driven by rapid human population growth and the resulting increase in demand for food and consumer goods, pushing many species to the brink of extinction. Forests, the most biodiverse habitats, have been the worst affected by this devastation. In Madagascar, for example, nearly 90 per cent of native forest cover has been lost, threatening many lemur species. Lemurs are found only here, and with the human population increasing, the lemurs' survival relies on preserving natural habitats. Even ring-tailed lemurs, one of the most widely recognized species, is threatened through habitat destruction. Ring-tailed lemurs breed well in captivity, and they are very adaptable, so reintroduction programmes are possible in areas where they have been lost – but success in this is dependent on them having a habitat to return to.

3-SECOND DISSECTION
Human populations are growing exponentially, putting enormous pressure on natural habitats and causing the decline and extinction of many species across the world.

3-MINUTE SYNTHESIS
As well as habitat loss, humans are degrading and fragmenting what remains of our natural areas. Improving connections between habitat patches and protecting existing areas for wildlife are top priorities. Some countries are currently attempting ambitious rewilding projects as well as schemes to reconnect the fragmented landscape, and many of those involved are local landowners who understand that reconnecting areas of conservation value will help many species and provide resilience in the face of climate change.

RELATED TOPICS
See also
PATRICIA WRIGHT
page 136

CLIMATE CHANGE
page 138

30-SECOND TEXT
Rebecca Thomas

Loss of native habitat directly affects a species' ability to survive through increased competition for declining resources, from food to places to hide.

POLLUTION

the 30-second zoology

3-SECOND DISSECTION
Pollution is a global problem: human activities introduce undesirable contaminants that cause harm to individuals, populations, ecosystems and even the whole ecosphere.

3-MINUTE SYNTHESIS
In 1945 the insecticide DDT became readily available, and its wide-scale use had devastating effects on wildlife. The chemical became embedded in the fat stores of many species. When they were eaten by their predators the toxin accumulated at increasing concentrations, and this bioaccumulation caused enormous harm to non-target species. Most notably, many birds of prey suffered, and it was only after DDT was banned that populations of species such as the peregrine falcon could begin to recover.

Almost every human activity causes some form of pollution, and this, in turn, causes enormous harm to the natural world as well as being responsible for many human deaths. Large-scale oil spills capture our attention. Thankfully they are rare, albeit devastating to local wildlife, but other forms of pollution are more hidden and insidious, damaging our climate, destroying the ozone layer, killing non-target insects, simplifying ecosystems and even affecting how animals develop. Recently awareness has risen of just how widespread plastic pollution is. Large plastics in the oceans are regularly ingested by marine mammals and birds. These are not digested and can prevent them from taking in food, meaning they will starve. Plastics also break down into tiny pieces becoming microplastics; these are ingested and incorporated into the tissues of many species, including the fish we eat. Microplastics have been found in every habitat in which they have been looked for, including the abyssal depths. They can be eaten by larval insects in rivers, and as the adults emerge they carry the microplastics and may be eaten by birds, passing the pollution along the food chain. Nowhere escapes the unwelcome touch of pollution, and this is the clearest sign that we have truly entered the Anthropocene.

RELATED TOPICS
See also
CLIMATE CHANGE
page 138

3-SECOND BIOGRAPHY
RACHEL LOUISE CARSON
1907–64
American biologist whose 1962 book *Silent Spring* brought about the environmental movement and led to the banning of DDT in many countries and the creation of the US Environmental Protection Agency

30-SECOND TEXT
Rebecca Thomas

Chemical contaminants, through accidental and deliberate acts, are not simply damaging to the environment. By entering the food chain pollutants threaten all organisms on the planet.

OVER-HARVESTING

the 30-second zoology

3-SECOND DISSECTION
Harvesting individuals at rates faster than their populations can recover leads to over-exploitation, and this is a major threat to many wild species.

3-MINUTE SYNTHESIS
Over-harvesting is a serious challenge in aquatic environments. Many fisheries across the world are a common resource and because of this suffer from the so-called 'tragedy of the commons'. No individual person fishing has the motivation to stop once a resource comes close to over-exploitation, because someone else will simply take the resource instead. This, alongside the technological developments that allow more fish to be taken with less effort, means that many fish stocks have collapsed.

Humans have always harvested wild animals and plants, but advances in technology and rising human populations now place immense pressures on many species. Many are under threat or have become extinct through over-exploitation – and not just for food and fur. Some are threatened by demand for trophies and traditional medicine – causing the over-harvesting of animals such as rhinoceroses, tigers and elephants – and unsustainable demand for bush meat from rapidly growing urban populations places a wide range of species in danger. Aquatic environments are perhaps the worst affected, with populations of many long-lived and highly valued fish, including cod and tuna, suffering enormously high losses. Harvesting can be sustainable if species can reproduce at a rate that compensates for mortality, but once a threshold is passed reproduction can no longer keep up with removal rates. These maximum sustainable yields are used in the fishing industry to estimate how many individuals can be safely harvested – which, for slowly reproducing species, can be a very small number – although some criticize this approach as failing to capture the complexity of pressures on fish populations. More importantly, it requires real political will to agree and enforce realistic quotas in international waters, something that has been sadly lacking to date.

RELATED TOPICS
See also
HUMAN–WILDLIFE CONFLICT
page 148

30-SECOND TEXT
Rebecca Thomas

Short-term gains are impacting on the ability of many species to recover their numbers, tipping them perilously close to extinction.

INVASIVE SPECIES

the 30-second zoology

3-SECOND DISSECTION
A number of the species we know and love are invasive, especially in more urbanized areas, but they can be a big driver of biodiversity loss.

3-MINUTE SYNTHESIS
Many species have been introduced outside of their native ranges, but this does not make them invasive. To get this title they need to cause harm, and only around 10 per cent of introductions become problematic. Unfortunately, some of the most damaging species – such as cane toads, cats, Nile perch and rosy wolf snails – have been released deliberately with dreadful ecological and economic consequences. Introducing species can result in biotic homogenization, reducing natural differences in biodiversity between environments.

Some species, through human interference, find themselves outside of their natural range. Some introductions – pet cats and house sparrows, for example – have been deliberate, but most, such as zebra mussels being transported across the oceans in the ballast water of ships, have been accidental. Many such species have little obvious effect on their new homes, but some spread in a way that greatly impacts native inhabitants, human health or the economy, and these are known as invasive species. Such invaders can be particularly damaging on islands. In the late 1940s the brown tree snake found its way to the Pacific island of Guam, most likely by stowing away on military aircraft. A lack of predators and competitors allowed them to thrive, and today they are devastating the island's ecology. It can be very difficult to remove invasive species once they become established, although there have been some success stories. Invasive rodents have been removed from Teuaua Island in French Polynesia, allowing the native sooty terns to nest on the island without the threat of rats predating their young. However, the cost of removing invasive species is often much greater than the cost of prevention, underlining the fact that it is better to avoid their introduction in the first place.

RELATED TOPICS
See also
HUMAN–WILDLIFE CONFLICT
page 148

URBAN ANIMALS
page 150

30-SECOND TEXT
Rebecca Thomas

The accidental or deliberate introduction of species to areas that lack the checks and balances of their natural range has resulted in some notorious losses of biodiversity.

HUMAN–WILDLIFE CONFLICT

the 30-second zoology

3-SECOND DISSECTION
Human–wildlife conflict
is increasing as human
populations grow and our
tolerance for crop damage,
disease and wildlife
interference diminishes;
because of this many
species face persecution.

3-MINUTE SYNTHESIS
Numerous strategies to
reduce human–wildlife
conflict have been used
when the species concerned
are of conservation value.
From lethal control to
compensation payments,
each aims to minimize the
contact between the parties
involved. Some of the most
successful examples have
been in providing
alternative livelihoods to
local people through
ecotourism. If people are
able to make their livings by
protecting the species that
would otherwise have been
causing problems then
communities are much more
accepting of their presence.

For many, a connection with nature is part of what makes life special and memorable. For these people, contact with wildlife is seen as a boon; holidays are focused on natural areas, wildlife documentaries are enjoyed and, even on our doorsteps, providing food for garden birds and other wildlife is something many people do. But when humans and wildlife mix conflicts can arise, with wild animals seen as a threat to livelihoods, health and well-being. For those who rely on agriculture, crop losses mean financial losses, and human–wildlife conflicts around farming can be devastating. Elephants epitomize this duality. For ecotourists, elephants are one of the most charismatic of mega-fauna, a species worth travelling around the world to see in the wild. But elephants can also cause extensive damage when raiding agricultural areas to find food, causing damage by trampling crops while moving through the fields. This can have serious consequences for the local human communities, and as a result persecution does take place. But working with local knowledge is often the best method for reducing conflict. Elephants do not like plants containing capsaicin, and they also have an aversion to bees, so farmers use chilli-plant fences and beehives to deter them.

RELATED TOPICS
See also
HABITAT LOSS
page 140

URBAN ANIMALS
page 150

30-SECOND TEXT
Rebecca Thomas

Sharing the planet with other species requires strategies that promote mutual harmony and sustainable solutions.

URBAN ANIMALS

the 30-second zoology

3-SECOND DISSECTION
Across the world cities are becoming bigger and more numerous, and many animals are learning to adapt and exploit the habitats and resources that they provide.

3-MINUTE SYNTHESIS
Speciation (the process by which genetically isolated animals evolve into distinct species) is happening within our own time. Mosquitoes first colonized the London Underground when the tunnels and stations were used as bomb shelters during the Blitz. Research during the 1990s revealed that the Underground mosquitoes had, by remaining in reproductive isolation, formed a subspecies, distinct even from those occupying other parts of the city, because they were unable to interbreed with other populations.

Humans have been modifying the landscape for thousands of years, but only within the past century have we been creating densely inhabited towns and cities. With over 50 per cent of the global human population now living in urban areas, many animals are learning to live alongside the city dwellers. Our built environments can be tough for animals – generally, as urbanization increases biodiversity is reduced – but some are finding ways to live, even thrive, among people. Mostly these are generalists such as rats and foxes, but alongside the ubiquitous sparrows and starlings we now find species as special as leopards and langurs, penguins and peregrines, and for some species urban areas are critical for their conservation. Under this novel evolutionary pressure animals are rapidly adapting genetically to urban life, especially in places where barriers restrict movement between populations – one perhaps surprising example of which is the London Underground mosquito *Culex pipiens molestus*. Many birds, too, show how species can change as they live alongside humans. Ornithologists have discovered that urban populations of the European blackbird sing at a higher pitch (because of traffic noise) and have shorter beaks (because food types differ) and no longer migrate during winter (because food is available all year round). Cities are the newest crucibles of evolutionary change.

RELATED TOPICS
See also
NATURAL SELECTION
page 18

HUMAN–WILDLIFE CONFLICT
page 148

3-SECOND BIOGRAPHIES
KATHARINE BYRNE
unknown

RICHARD NICHOLS
1959–
Researchers who undertook a study of London Underground mosquitoes in the 1990s

30-SECOND TEXT
Rebecca Thomas

As ever-increasing numbers of us become urban dwellers, our interactions with other species are being shaped by those animals that are able to live alongside us. Losing connection with nature has implications for conservation world wide.

APPENDICES

RESOURCES

BOOKS AND ARTICLES

Amphibian (Eyewitness series)
Barry Clarke
(Dorling Kindersley, 2005)

Animal Fact File: Head-To-Tail Profiles of Over 90 Mammals
Tony Hare
(Facts on File Inc., 1999)

'*Culex pipiens* in London Underground tunnels: Differentiation between surface and subterranean populations'
Katharine Byrne and Richard A. Nichols
Heredity 82 (Pt 1)(1):7–15, February 1999
DOI: 10.1038/sj.hdy.6884120

The Diversity of Life
Edward O. Wilson
(Penguin, 2001)

Extraordinary Birds: Exquisite Selections of Art and Ornithology from the American Museum of Natural History Library, Essays & Plates
Paul Sweet, including a chapter by Peter Capainolo
(Sterling Signature, 2013)

Fishes of the World
Joseph S. Nelson
(John Wiley & Sons, 2006, 4th edn)

The Great Big Book of Snakes and Reptiles
Barbara Taylor
(Hermes House, 2006)

Herpetology, An Introductory Biology of Amphibians and Reptiles
Laurie J. Vitt and Janalee P. Caldwell
(Elsevier Science, 2013)

Insect Evolutionary Ecology
M. D. E. Fellowes, G. J. Holloway and J. Rolff
(CABI Publishing, 2005)

Mammals
Juliet Clutton-Brock
(Dorling Kindersley/Smithsonian Handbooks series, 2002)

Ornithology
Frank B. Gill
(W. H. Freeman & Co., 2006, 3rd edn)

Ornithology in Laboratory and Field
Olin Sewall Pettingill
(Academic Press, 1985, 5th edn)

Reptile (Eyewitness series)
(Dorling Kindersley, 2005)

Sharks of the World
Leonard Compagno, Marc Dando and Sarah Fowler
(Princeton University Press, 2005)

Silent Spring
Rachel Carson
(Houghton Mifflin, 2002, anniversary edn; first published in 1962)

Song of the Dodo: Island Biogeography in an Age of Extinction
David Quammen
(Simon & Schuster, 1997)

Vertebrate Life
F. Harvey Pough and Christine M. Janis
(Oxford University Press, 2018, 10th edn)

When Life Nearly Died: The Greatest Mass Extinction of All Time
Michael Benton
(Thames & Hudson, 2015)

Wilding: The Return of Nature to a British Farm
Isabella Tree
(Picador, 2019)

Wonderful Life: Burgess Shale and the Nature of History
Stephen Jay Gould
(Vintage, 2000)

ONLINE READING

All the World's Primates
Fully referenced online resource covering all 505 species of primates
www.alltheworldsprimates.org/Home.aspx

State of the World's Birds
Report by Birdlife International using birds to assess the condition of global ecosystems.
www.birdlife.org/sowb2018

Edge of Existence
Listing the most threatened animals on Earth, taking into account both rarity and taxonomic isolation, this is a sobering account of species on the brink of extinction provided by the Zoological Society of London.
www.edgeofexistence.org/species/

The Global Biodiversity Information Facility
An astounding international network and research database providing open access to information about all types of life on Earth.
www.gbif.org

NOTES ON CONTRIBUTORS

EDITOR

Mark Fellowes has been passionate about wildlife since an early age, spending his formative years in the wilds of the west of Ireland surveying birds and bringing bits of the countryside back home. He completed his BSc in zoology and a PhD in evolutionary biology at Imperial College London. Following a brief post-doctoral stint at the NERC Centre for Population Biology at Imperial, he joined Reading as a lecturer. Since then he became Professor of Ecology, and he is now Pro-Vice Chancellor at the University. Professor Fellowes has published several books and numerous papers in research journals. His current work focuses on interactions between people and wildlife, with an emphasis on urban ecosystems. He works on insects, birds and mammals, with ongoing projects in Ghana, Nigeria, India, the USA and Brazil, as well as in the UK. He undertakes a range of outreach work to engage the public and media in the importance of biodiversity and was listed as one of the UK's 100 most influential men under the age of 40 by *Esquire* magazine.

CONTRIBUTORS

James Barnett is a behavioural ecologist based at McMaster University, Ontario, Canada. He has an MSc and a PhD from the University of Bristol, UK, and his postdoctoral research includes the study of predator–prey interactions and the way animal colours function, evolve and are perceived by different species.

Amanda Callaghan is Professor of Invertebrate Zoology at the University of Reading, UK, and Curator of the Cole Museum of Zoology located on the university's campus. Her research interests include mosquito ecology, entomology and the effect of microplastics and ecotoxicology in freshwater invertebrates. She is a Fellow of the Royal Society of Biology, and the Royal Entomological Society. A former Editor of *Antenna*, the bulletin of the RES, Professor Callaghan is also the author of numerous papers and articles for academic journals.

Peter Capainolo has had an interest in natural history, particularly ornithology, since boyhood. Birds of prey fascinated him, and he received one of the first falconry licences issued by New York State at the age of 18. He studied zoology and practised falconry under Professor Heinz Meng at the State University of New York at New Paltz. Currently he is Senior Scientific Assistant/Senior Museum Specialist in the Division of Vertebrate Zoology at the American Museum of Natural History and an adjunct faculty member in the Department of Biology at The City College of the City University of New York. Peter is the author of a book on birds of prey and has contributed to several books on ornithology in general.

Neil Gostling is an evolutionary biologist and a Senior Teaching Fellow at the University of Southampton, UK. Originally from London, he spent every opportunity of his childhood at the Natural History Museum, and watching David Attenborough documentaries. He received a BSc (botany and zoology) and a PhD (evolutionary developmental biology) from the University of Reading, before moving to Bristol to study fossil embryos from China, and the Cambrian period. His research has looked at the evolution of some of the earliest birds, dinosaurs and mammals. Neil runs the MRes in evolution and annually leads a field course to the Galápagos. He regularly gives talks to natural history and Café Sci groups on Darwin, evolution and biology.

Rebecca Thomas is an urban ecologist and Senior Teaching Fellow at Royal Holloway, University of London, UK. Her research interests focus on the conservation biology and ecology of birds and mammals, specifically in how people's decisions affect the ecology of wild species. Dr Thomas is particularly interested in how human activity affects ecological interactions at a range of scales, and especially in trying to uncover some of the unexpected and unpredicted consequences of our behaviour.

INDEX

ACKNOWLEDGEMENTS

The publisher would like to thank the following for permission to reproduce copyright material on the following pages:

All images that appear in the montages are from Shutterstock, Inc. unless stated.

Donald Griffin image used courtesy of the Rockefeller Archive Center.

Mary Jane West Eberhard image used with kind permission of Mary Jane West Eberhard.

Alamy/ 19th era 2: 39, 83; Age Fotostock: 41; Andrey Nekrasov: 57; Ann and Steve Toon: 95; Antiqua Print Gallery: 79, 139; Anton Sorokin: 125; Archivist: 103; Arterra Picture Library: 145; Bazzano Photography: 83; blickwinkel: 79, 139; Bookworm Classics: 43; BSIP SA: 41; Everett Collection Inc: 143; Florlleglus: 23, 79, 119; FLPA: 45; Gaertner: 129; gameover: 41, 109; Geerati Nilkaew: 99; Glasshouse Images: 123; Henri Koskinen: 41; Historic Collection: 67, 77, 125; Historic Images: 77; Image BROKER: 139; INTERFOTO: 39; Joel Sartore: 147; Kiyoshi Takahase Segundo: 17; Library Book Collection: 41, 45, 57, 79; Mark Conlin: 145; Maxim Tatarinov: 127; Michael Rhys Williams: 37; Narelle Power: 147; Natural History Museum: 19, 57; Nature Photographers Ltd: 45, 57; Nikolay Staykov: 77; nobleIMAGES: 149; Old Images: 17, 103; Oldtime: 109; Opera Nicolae: 19; Patrick Guenette: 17; Petr Pavluvcik: 119; RGB Ventures/SuperStock: 147; Richard Mittleman/Gon2Foto: 129; Rob Cousins: 115; Science History Images: 23, 57, 75, 147, 151; Steve Taylor ARPS: 149; Stuart Donaldson: 47; The Book Worm: 57; The History Collection: 99; Tot Collection: 125; WaterFrame: 45; WILDLIFE GmbH: 123; World History Archive: 27; www.pqpictures.co.uk: 43.

Biodiversity Heritage Library: 41, 67, 100, 103.

Boston Public Library: 19.

Flickr/ British Library: 67; Biodiversity Heritage Library: 99.

Getty/ Mike Windle: 137; Nnehring: 43, 45; ZU_09: 43.

Internet Archive/ Biodiversity Heritage Library: 25, 97, 105, 117; Smithsonian Libraries: 85.

Wikimedia Commons/ Audubon: 131; Biodiversity Heritage Library: 45, 67; Bonhams: 105; British Library/Flickr Commons: 131; Bstelnitz: 55; Colour Illustration by Mrs. Sarah L. Martin / University of Texas at Austin, School of Biological Sciences: 107; E.J. Spitta: 47; Esculapio: 127; Internet Archive: 67, 107; Internet Archive/NCSU Libraries: 47; Internet Archive/Open Knowledge Commons and Harvard Medical School: 75; Kunstformen der Natur / Padent~commonswiki: 85; Internet Archive/Book Images: 21; James St. John: 21; Library of Tasmania: 25;Minneapolis Institute of the Arts: 99; NOAA: 37; Nobu Tamura: 25; PLOS: 90; Proceedings of the Zoological Society of London: 151; Rhino Resource Center/Hermann Schlegel: 127; Siga: 47; Smithsonian Institution Archives: 21; Southeby's: 67; Wellcome Images: 75; художник: 14.

All reasonable efforts have been made to trace copyright holders and to obtain their permission for the use of copyright material. The publisher apologizes for any errors or omissions in the list above and will gratefully incorporate any corrections in future reprints if notified.